PENGUIN

MATHS

Gaurav Tekriwal is the founder president of the Vedic Maths Forum India. An educator, Gaurav has been imparting high-speed Vedic mental-mathematics skills over the past fifteen years across the globe. He inspires and informs people, helping them to realize their true potential by introducing them to the world's fastest mental-maths system—Vedic mathematics.

Gaurav is the author of *Speed Math* and his complete DVD set on the topic is popular among students and academicians worldwide.

Through television programmes, DVDs, books, workshops and seminars, Gaurav has taken the Vedic maths system to over four million students in India, South Africa, Australia, the United States and Oman. He is a four-time TED speaker, and has been recently awarded the INDIAFRICA Young Visionaries Fellowship by the ministry of external affairs, India.

For more information, please visit www.mathssutra.com.

MATHS SUTRA

The Art of Vedic Speed Calculation

GAURAV TEKRIWAL

PENGUIN BOOKS

An imprint of Penguin Random House

PENGUIN BOOKS

USA | Canada | UK | Ireland | Australia
New Zealand | India | South Africa | China

Penguin Books is part of the Penguin Random House group of companies
whose addresses can be found at global.penguinrandomhouse.com

Published by Penguin Random House India Pvt. Ltd
7th Floor, Infinity Tower C, DLF Cyber City,
Gurgaon 122 002, Haryana, India

Penguin
Random House
India

First published by Penguin Books India 2015

ISBN 9780143425021

Typeset in Electra LT Std by R. Ajith Kumar, New Delhi
Printed at Repro Knowledgecast Limited, India

www.penguin.co.in

To my mother, Madhu Tekriwal, for her unconditional love, support, encouragement and guidance

Contents

Preface

INDIA IS A LAND with a golden heritage. It has given to the world jewels such as yoga (which is now a multibillion-dollar industry), Ayurveda (another multibillion-dollar industry) and chicken tikka (a trillion-dollar taste!).

Indians are feted worldwide for their outstanding cricket and maths skills; after all, it was we who discovered Sachin, the zero and the decimal system.

This book is about being the Sachin of the decimal system and scoring high in a limited-time competitive examination through the use of maths sutras.

Maths sutras are the secret with which ancient Indians performed their feats over 5000 years ago. This system was lost to the world until it was discovered by a scholar in the forests of south India, some fifty years ago; so it can be labelled as a twentieth-century phenomenon.

These maths sutras hold the power to speed up your calculations, give you confidence, and make maths fun and interesting. They will help you in studies just the same way the yoga sutras of Patanjali help improve one's health and the ancient wisdom of Vātsyāna's *Kama Sutra* helps one in the field of love.

So what are these sutras? Where do they actually come from? Let's explore that a bit further.

We all know that the Indian civilization is one of the oldest in the world. The Indian sages passed on their collected works orally from one generation to the other using codes which unlock various layers of meaning. They compiled four texts in Sanskrit known as the Vedas—meaning 'knowledge'.

The *Rig Veda*, the *Yajur Veda* and the *Sama Veda* contain various hymns dedicated to nature gods, whereas the fourth or the *Atharva Veda* contains a collection of spells, incantations and speculative hymns. It is in these ancient texts that the maths sutras are found in coded form, according to the founder, Tirthaji.

Consider this song dedicated to Lord Krishna:

Gopi bhagya madhuvrata
srngiso dadhi sandhiga
khala jivita khatava
gala hala rasandara

The literal translation is as follows: 'O Lord anointed with the yoghurt of the milkmaids' worship (Krishna) / O saviour of the fallen / O master of Shiva / please protect me.'

But when you apply the numerical code and the master key to it you get the value of pi to 32 places of decimal.

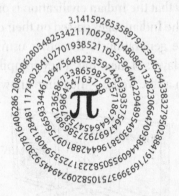

Unbelievable, isn't it? Such is the power of the maths sutras, also popularly known as Vedic mathematics.

Sutras are original thoughts, spoken or written in a concise and memorable form. Maths sutras are basically short mathematical formulas dealing with arithmetic and algebra. Sixteen of these were rediscovered by Tirthaji.

But in this age of technology how relevant are these sutras?

With each passing year the scope of these maths sutras gets wider, as globally more and more students deal with maths. This has caused a full-grown numeracy crisis. Here I present some surprising data:

75.2 per cent of all the children in grade five in India can't do division (three-digits-by-one-digit problems).
Source: ASER 2012 by Pratham

73.7 per cent of all children in grade three in India can't subtract (two-digit problems with borrowing).
Source: ASER 2012 by Pratham

Almost half of British adult population have the maths skills of an eleven-year-old or less. That's 17 million adults in the UK alone.
Source: Telegraph, UK

46.3 per cent is the maths pass rate in South Africa's national senior certificate examinations. Maths is a national disaster in the country.

Such is the havoc that maths has created across countries. Most children across the world can't do maths.

This is where Vedic maths comes in with its sheer simplicity and comfort, and makes a career possible in the fields of management, engineering, banking, finance and law by helping to crack their respective examinations.

The maths sutras open up a world of possibilities for everyone. Problems which once seemed difficult and daunting are child's play now. The sutras show that, although

maths problems may seem abstract and unbelievable in the beginning, they all have principles of logic behind them.

Most of the sutras are very visual when you apply them. Take the 'vertically and crosswise' sutra, for example. The pattern can be extended to do large multiplications and even shortened to do smaller two-digit-by-two-digit sums. This visual pattern makes the sutras very easy to remember and practise.

I've had a fascinating journey in the field of Vedic mathematics over the last fifteen years. This journey has not only introduced me to some very interesting people but also allowed me to travel the length and breadth of the world — across the seven seas. And wherever I went, right from meeting kids in the slums in Dharavi in Mumbai to the grand casinos of Cape Town in South Africa to Wall Street in New York, I saw there was one thread binding them all — Vedic mathematics.

To a kid in Dharavi, Vedic maths is like a survival skill, because the child is expected to earn a livelihood at a very early age. Similarly, casino dealers have to be quick with calculations for obvious reasons, and the stock market people need it for split-second decisions.

People hate maths for one big reason—lack of proper communication. If mathematical principles are communicated properly and the concepts taught with passion, there would be no phobia about maths and it would even be fun and interesting! This book is one step in that direction and all the ideas have been communicated clearly with enough examples to practise.

This book is divided into nineteen chapters and each chapter deals with a different topic, attempting to cover the concepts as well as their applications. Each topic is independent and can be studied in isolation, without having to read the entire book, thereby making it an easy reference manual.

May the power and influence of the maths sutras be upon you, and may you secure your dream seat at your preferred institution!

So without further ado, let's get started.

Multiplication

IN JANUARY 2015, Neha Manglik, a chemical engineer from BITS Pilani, made headlines. She was the only woman to score 100 percentile in the Common Admission Test (CAT) 2014. In her first interview with *India Today*, she admits that Vedic maths helped her to score better!

Such a thing was unheard of until then. What Neha had achieved by using the maths sutras is the cherished dream of lakhs of students in India. Vedic maths helps you realize your dream and that's why these maths sutras are becoming popular.

Let us now try to understand one of the most popular sutras on multiplication—the Base Method of Multiplication.

In Vedic maths, multiplication methods are of two kinds: one which can be applied only to special cases and another which can be applied to all the cases. The first kind is important because the methods are much simpler and quicker; the second one is important because you can plug them in where the first doesn't work. We will explore both types of multiplication methods in this chapter.

The Base Method of Multiplication

This method of multiplication is used to multiply numbers very close to powers of 10, like 10, 100, 1000, 10000 and so on.

In the beginning, we learn the fundamentals, so we stick with single-digit figures; but as soon as we master multiplying single-digit figures we will move on to solving double-digit and triple-digit figures as well.

So let's kick off with the sum 9 x 8. Now we all know that the answer is 72. It's not rocket science. But, as I said, we will first learn the fundamentals with single-digit figures before moving on to bigger digits. So please bear with me.

Step 1

The first thing to note is that both the numbers are very close to 10. So our base becomes 10.

9 is less than 10 by 1. So we write -1 on the right-hand side. Similarly, for 8 we write -2 because 8 is less than 10 by 2.

```
      9 | - 1
  x   8 | - 2
  _____|_____
         |
```

Step 2

Now that the sum is laid out, we apply our first rule which is: we **add or subtract crosswise as the sign may be**. Therefore we subtract here because the sign is a minus. So we do 9 - 2 = 7. We may also do 8 - 1 = 7. Both give us the first left-hand digit 7.

```
      9  - 1
  x   8  - 2
  _____|_____
      7  |
```

TIP: Here we have a choice to either do 9 - 2 or 8 - 1 = 7, whichever is easier. Take your call.

Step 3

In our final step, we apply the second rule: **multiply vertically on the right-hand side**. So here we multiply -1 and -2, which gives us +2. We then place 2 in the unit's column and we have our final answer 72.

$$\begin{array}{c|c} 9 & -1 \\ \times\ 8 & -2 \\ \hline 7 & 2 \end{array}$$

Let us now try to solve another simple sum to help get familiar with the steps.

Our sum now is 8 x 7. We lay out the digits exactly as in the last sum.

$$\begin{array}{c|c} 8 & \\ \times\ 7 & \\ \hline & \end{array}$$

Step 1

Now we take our base as 10 as both the numbers are near to ten. 8 is less than 10 by 2, so we write -2 on the right-hand side. Similarly, 7 is less than 10 by 3, so we put -3 on the right-hand side of it.

$$\begin{array}{c|c} 8 & -2 \\ \times\ 7 & -3 \\ \hline & \end{array}$$

Step 2

Now applying the first rule, we add or subtract crosswise as the sign may be. 8 - 3 is 5. We could also do 7 - 2, giving us 5 as the first digit of the answer.

$$
\begin{array}{c|c}
8 & -2 \\
\times\ 7 & -3 \\
\hline
5 &
\end{array}
$$

Step 3

Finally, we multiply vertically on the right-hand side. So -2 multiplied by -3 give us +6. We put down 6 in the unit's place, giving us the final answer, 56.

$$
\begin{array}{c|c}
8 & -2 \\
\times\ 7 & -3 \\
\hline
5 & 6
\end{array}
$$

We now take an important single-digit multiplication problem. Say we have to multiply 7 x 6. Here's how you do it:

We first lay out the sum — just the way we have learnt now.

$$
\begin{array}{c|c}
7 & \\
\times\ 6 & \\
\hline
&
\end{array}
$$

Step 1

Both the numbers are close to the base of 10. 7 is less than 10 by 3. And 6 is less than 10 by 4. Therefore, we write -3 and -4 at their relevant positions.

$$
\begin{array}{c|c}
7 & -3 \\
\times\ 6 & -4 \\
\hline
&
\end{array}
$$

Step 2

We now apply the first rule: **add or subtract crosswise as the sign may be**. 7 - 4 is 3. We could also do 6 - 3, giving us 3 as the first digit of our answer.

$$
\begin{array}{r}
7 \quad -3 \\
\times\ 6 \quad -4 \\
\hline
3
\end{array}
$$

Step 3

We multiply vertically on the right-hand side of the problem. -3 times -4 equals 12. Now here we apply our important **placement rule**, because we have two digits. **The placement rule says that if the base has 'n' number of zeroes, there will be *n* number of digits on the right-hand side**. In this case, it means that, since our base 10 has one zero, there will be only one digit on the right-hand side. Any extra digits will be carried forward.

So here we have 12 on the right-hand side. We let 2 remain in the unit's place and carry over 1 to the left and add to 3. 3 + 1 = 4, which becomes our ten's place. So our final answer is 42.

$$
\begin{array}{r}
7 \quad -3 \\
\times\ 6 \quad -4 \\
\hline
3 \quad {}_1 2
\end{array}
$$

42

NOTE: This placement rule can be applied to all other bases like 100, 1000, 10000, etc., as we will see.

I was in South Africa a couple of years ago doing research on mathematics. I asked a fourteen-year-old boy how much 8 times 7 is. He looked at me, then opened his exercise book and started drawing circles. I noticed that he drew eight circles and then again eight circles till he drew seven such sets—56 circles. Then he started counting them and finally gave me the wrong answer of 54.

I wondered what method had he used and why.

After asking a couple of teachers, I found out that South Africa as a country is struggling with the way mathematics is taught. There are loads of reasons for this. So I decided I would teach this teen the base method. And you won't believe the reaction on the face of this youngster when I showed him this method! Then he asked me if the method works with other numbers as well. And I said, 'Most definitely,' and showed him too.

He was completely taken aback when I showed him how

to do quick multiplications with larger numbers like this one: 99 x 97.

So now let us take this sum, 99 x 97. Here the base is 100 as both 99 and 97 are near 100. So let us lay out the sum as before:

$$\begin{array}{r|l} 99 \\ \text{x } 97 \end{array}$$

Step 1

99 is less than our base 100 by 1. So we write -01 on the right-hand side. Similarly, 97 is less than the base 100 by 3, so we write -03 on the right-hand side, below -01. Also note that we write -01 and -03, because of the base which is 100, and 100 has two zeroes. So, on the right-hand side, there should be two digits.

$$\begin{array}{r|l} 99 & -01 \\ \text{x } 97 & -03 \end{array}$$

Step 2

We now cross-add or subtract as the sign may be. So here we cross subtract 99 - 03 or 97 - 01 giving us 96, the first part of the answer.

$$\begin{array}{r|l} 99 & -01 \\ \text{x } 97 & -03 \\ \hline 96 \end{array}$$

Step 3

In our final step, we multiply vertically. We multiply -01 and -03, which gives us 03. So our answer is 9603. It is important to add a zero before 3, because of the placement rule. Since our base is 100, there should be two digits on the right-hand side. So our answer is 9603.

```
      99 | -01 ↑
    x 97 |✗-03 |
      96 | 03
```

Let us now take another sum and crack it using the same method. Say, for example, we have 96 x 98.

We lay the sum out as before:

```
      96 |
    x 98 |
---------+
         |
```

Step 1

```
      96 | -04
    x 98 | -02
```

Step 2

```
      96 | -04
    x 98 |✗-02
---------+
      94 |
```

Step 3

```
      96 | -04 ↑
    x 98 |✗-02 |
---------+
      94 | 08
```

Till now the sums were pretty straightforward. Let's take a sum which has carry-overs. For example, let's take 89 x 88.

We lay the sum down as before:

Step 1

$$\begin{array}{r|r} 89 & -11 \\ \times\ 88 & -12 \\ \hline & \end{array}$$

Step 2

$$\begin{array}{r|r} 89 & -11 \\ \times\ 88 & -12 \\ \hline 77 & \end{array}$$

Step 3

$$\begin{array}{r|r} 89 & -11 \\ \times\ 88 & -12 \\ \hline 77 & 32 \end{array}$$

7832

TIP: You can multiply 11 and 12 mentally by the 'multiplication by 11' technique. Eleven times 12 is 132.

Now that we have got the hang of this system, let's use numbers below the base of 1000.

Say we have 998 x 997.
We lay out the sum just as before.

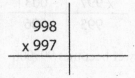

Step 1
Here the base is 1000. So we can take the difference of 998
and 997 from 1000. We get -002 and -003 respectively. This
we put at the appropriate places.

$$\begin{array}{c|c} 998 & -002 \\ \times\,997 & -003 \end{array}$$

Step 2
We now cross-subtract. So we do 998 - 003 = 995. We can
also do 997 - 002 = 995. Both ways give us 995. We put
down 995 as the first part of the answer.

$$\begin{array}{c|c} 998 & -002 \\ \times\,997 & -003 \\ \hline 995 & \end{array}$$

Step 3
We now multiply vertically. Multiplying -002 and -003,
we get 006. We put down 006 as our second answer digit.
The point to note here is that we give importance to the
placement rule. We don't write simply 6 but we write 006,
as the base 1000 has three zeroes.

```
998        - 002
x 997      - 003
─────     ──────
995        006
```

So our final answer becomes 995006.

This time let's take another sum. Say, for example, 123 x 999.

```
   123
 x 999
────────
```

Step 1

Here we see that we have to subtract 123 from 1000. From experience, I can safely say that a lot of you will have problems doing that mentally. At this point, to simplify things, I will introduce a maths sutra called 'All from 9 and last from 10'.

This sutra simply states that **when we have to subtract any number from powers of ten, we subtract all the numbers from 9 and the last from 10.** So in the case of 123, we subtract 1 and 2 from 9 and, the last digit 3 from 10. So we have

9 - 1 = 8 and 9 - 2 = 7 and 10 - 3 = 7. We also do 1000 - 999 = 1, so in the sum we also have -001.
So our sum now looks like this:

```
   123     - 877
 x 999     - 001
────────  ──────
```

Step 2

We now cross-subtract. 123 - 001 = 122 or 999 - 877 = 122.

Our sum looks like this now:

$$
\begin{array}{r|r}
123 \diagdown - 877 \\
\text{x } 999 \mid - 001 \\
\hline
122 \mid \\
\end{array}
$$

Step 3

In our final step, we multiply vertically on the right side.

So -877 x -001 = 877.

So our answer becomes 122877.

$$
\begin{array}{r|r}
123 \diagdown - 877 \\
\text{x } 999 \mid - 001 \\
\hline
122 \mid 877 \\
\end{array}
$$

Let's take a sum which has all the attributes of being a difficult sum.

So say we have 1234 x 9999. Now with the maths sutra 'All from 9 and last from 10' in place, this sum will seem like child's play! So let's crack it.

Step 1

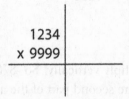

1234
x 9999

Upon seeing the sum we immediately apply the sutra 'All from 9 and last from 10'. So here the base is 10000. So 10000 - 1234 = 8766.

Here's how we get to 8766.

9 - 1 = 8,

9 - 2 = 7,

9 - 3 = 6 and finally,

10 - 4 = 6.

Also for 9999, we get the deficiency of -0001.

Our sum looks like this now:

$$
\begin{array}{r|r}
1234 & -8766 \\
\times\ 9999 & -0001 \\
\hline
\end{array}
$$

Step 2

We cross-subtract and get to the first part of our answer: 1234 - 0001 = 1233 or 9999 - 8766 = 1233. So 1233 is the first part of our answer.

$$
\begin{array}{r|r}
1234 & -8766 \\
\times\ 9999 & -0001 \\
\hline
1233 & \\
\end{array}
$$

Step 3

Finally, we multiply vertically! So -8766 x -0001 = 8766. This becomes the second part of the answer. Combining these we get the final answer 12338766.

$$\begin{array}{c|c}
1234 & -8766 \\
\times\ 9999 & -0001 \\
\hline
1233 & 8766
\end{array}$$

Above the Base

So far we have been doing multiplications of numbers which are below the base of 10, 100, 1000, etc. Now we will multiply numbers which are above these bases.

When I had the opportunity to speak on Vedic maths at TED Youth@New York, I chose the base method as it was by far the most engaging of all the methods, since it could be done mentally. Students could relate to it as, by applying this method, they could just shout aloud the squares of numbers more than 100 or 1000. And it was nothing but just understanding a simple pattern which I will share with you here.

Let's take a sum of, say, 12 x 13.

We lay it out the same way as before.

$$\begin{array}{r|} 12 \\ x\ 13 \\ \hline \end{array}$$

Step 1

Here we write down the excess from the base instead of the deficiencies. So 12 is more than the base 10 by 2, so we write +2 and similarly for 13, we write +3, because 13 is more than 10 by 3.

$$\begin{array}{r|l} 12 & +2 \\ x\ 13 & +3 \\ \hline \end{array}$$

Step 2

Now here we cross-add instead of cross-subtract as the numbers are above the base. So we have 12 + 3 = 15, which is the first part of our answer. We could also do 13 + 2 to arrive at 15.

$$\begin{array}{r|l} 12 & +2 \\ x\ 13 & +3 \\ \hline 15 & \end{array}$$

Step 3

Now we multiply vertically, just as before. +2 x +3 equals +6, which is the final part of our answer. So our answer becomes 156.

```
    12   +2
  x 13   +3
    15 | 6
```

This I think is pretty straightforward and can be done in one line mentally. Let's take a carry-over sum, before moving on to doing squares of numbers above 100 or 1000.

Say we have 18 x 19. We lay the sum down just as before, like this:

```
    18 |
  x 19 |
```

Step 1

The base here is 10. So we write the excess from 10 on the right-hand side. For 18, we write +8 and for 19, we write +9.

```
    18 | +8
  x 19 | +9
```

Step 2

We now cross-add and arrive at the first answer digit. 18 + 9 = 27. We could also do 19 + 8 = 27.

```
    18 | +8
  x 19 | +9
    27 |
```

Step 3

We now multiply vertically on the right-hand side. Eight times 9 is 72. We now apply the placement rule. Since our base is 10 and 10 has only one zero, there should only be one digit on the right-hand side. Any other digit we add by carrying it over to the left side.

So here we take 7 to the left-hand side and add it to 27. 27 + 7 = 34. And then 2 comes down, giving us the answer 342.

$$
\begin{array}{r}
18 \quad +8 \\
\times \quad 19 \quad +9 \\
\hline
27 \quad {}_{7}2
\end{array}
$$

342

Now we come to the section you will absolutely love. We take the base 100 here. I will share with you the pattern and, once you understand that, you will be able to work out squares more than 100, mentally.

Let's take the sum 101 x 101. Here we lay out the sum just as before.

$$
\begin{array}{r}
101 \\
\times \ 101
\end{array}
\Big|
$$

In this sum +01 is the excess from 100. So we write that on the right-hand side.

Now try doing the sum mentally.

Step 1

Imagine you are now cross-adding. 101 + 01 = 102. This gives us the first part of the answer.

$$\begin{array}{r|r} 101 \times \!\!\!\!\! & +01 \\ \times\ 101 & +01 \\ \hline 102 & \end{array}$$

Step 2

In this step, we multiply vertically, 01 by 01. This is nothing but squaring the excess of 01. So 01 squared gives us 01, which is the second part of our answer. So our complete answer is 10201.

$$\begin{array}{r|r} 101 \times \!\!\!\!\! & +01 \\ \times\ 101 & +01 \\ \hline 102 & 01 \end{array}$$

Now you can do this mentally. Just find out the excess and add it to the number which you are squaring. This will give you the first part of the answer.

Now to get the final part of the answer, just square the excess. And you have the answer.

Let's take another sum: 102 x 102.

Try doing this sum absolutely mentally. Find out the excess and add it to the number that is being squared. Finally, square the excess and you have the answer!

$$\begin{array}{r|r} 102 \times \!\!\!\!\! & +02 \\ \times\ 102 & +02 \\ \hline 104 & 04 \end{array}$$

Now say we have 103 x 103. You try it first mentally and arrive at the answer. Since you are only learning now, it's okay if you don't get it right. See the solution only after you have got an answer.

So in the sum 103 x 103, find out the excess which is +03.

Add 03 to 103. It gives us 106, which is the first part of the answer.

To get the second part of the answer, simply square 03. That's 09.

Combining this with the first part of the answer we get 10609, which is our final answer.

$$
\begin{array}{r|l}
103 \!\!\!\nearrow\!\!\! & +03 \\
\times\ 103 & +03 \\
\hline
106 & 09
\end{array}
$$

Now try 108^2 in the same way. Your answer would be 11664.

Remember there will be sums which will have carry-overs too. Say, for example, 111 x 111.

$$
\begin{array}{r|l}
111 \!\!\!\nearrow\!\!\! & +11 \\
\times\ 111 & +11 \\
\hline
122 & ,21
\end{array}
$$

12321

In this sum 111 x 111, the excess from the base of 100 is +11, which we write at the appropriate place. We then add the excess to 111, which gives us 122 as the first part of the answer.

Finally, we square the excess of 11 giving us 121. Here since the base is 100 and it has two zeroes, we will take 1 as the carry-over. We add 1 to 122, giving us 123 and our answer becomes 12321.

We will do one more similar sum. Say, we have 112^2.

Step 1

The excess is +12. We add +12 to 112 giving us 124.

$$112 \underset{\times}{} \begin{array}{c} + 12 \\ \end{array}$$

```
   112 ⟋+ 12
       ⟍
 x 112 | + 12
 ──────┼──────
   124 |
```

Step 2

Finally, we square the excess of 12. 12 x 12 = 144. We apply the placement rule and carry 1 to the left-hand side, adding to 124, and we get 125.

```
   112 ⟋+ 12
       ⟍
 x 112 | + 12
 ──────┼──────
   124 ←₁44
 ──────────────
     12544
```

So our final answer becomes 12544.

This was till base 100. Now let us see base 1000.
Say we have the sum 1006 x 1009.

Step 1

We lay the sum out as before:

```
   1006  |
 x 1009  |
 ────────┼
         |
```

Step 2

The base is 1000 and the excess from 1000 is +006 and +009 respectively. We do cross-addition and get the first part of our answer. 1006 + 009 = 1015, or 1009 + 006 = 1015.

$$\begin{array}{r|l}
1006 \diagdown +006 \\
\times\ 1009 \mid +009 \\
\hline
1015 \mid
\end{array}$$

Step 3

We multiply vertically now. So 006 x 009 gives us 054, which is the second part of our answer. Our complete answer is 1015054.

$$\begin{array}{r|l}
1006 \diagdown +006 \uparrow \\
\times\ 1009 \mid +009 \mid \\
\hline
1015 \mid 054
\end{array}$$

Now that we know how a base 1000 sum looks, we can even start squaring base 1000 problems with the same reasoning as the base 100 sums.

So for example, let's take the sum 1008 x 1008.

Try doing this sum mentally on the same logic principle shared earlier.

The excess is +008, which is added to 1008 to give us 1016, which becomes the first part of the answer.

To arrive at the second part of the answer, we simply square the excess. So 008 squared is 064.

Combining the two parts, our answer becomes 1016064.

$$\begin{array}{r|l}
1008 \diagdown +008 \uparrow \\
\times\ 1008 \mid +008 \mid \\
\hline
1016 \mid 064
\end{array}$$

Let us now see the principle behind this maths sutra and why it works.

$$(x + a)(x + b) = x(x + a + b) + ab$$

Here x is the base of 10, 100 or 1000, etc., and a and b are the excesses or deficiencies from the base.

Above and Below the Base

Now that we have seen the cases of numbers being below the base and above the base, we can move on to instances where one number is below the base and the other number is above the base.

Say, for example, we have 15 x 8.

We lay the sum out just as before:

$$\begin{array}{r} 15 \\ \times\ 8 \\ \hline \end{array}$$

Step 1
Both 15 and 8 are above and below the base of 10. 15 is above 10 by 5, so we write +5 and 8 is below 10 by 2, so we write -2.

$$\begin{array}{r|l} 15 & +5 \\ \times\ 8 & -2 \\ \hline \end{array}$$

Step 2

We cross-add or subtract as the case may be over here. So we do 15 - 2 = 13. We can also do 8 + 5, which equals 13 as well.

We then multiply vertically +5 and -2 to get -10 on the right-hand side. Our sum now looks like this:

$$
\begin{array}{r|r}
15 \diagdown +5 \\
\times\ 8 \diagup -2 \\
\hline
13 & -10
\end{array}
$$

Step 3

Since we have -10 on the right-hand side, there is no need to panic! What we need to do is multiply the left-hand side (13) by the base (10). Thirteen times 10 is 130, and from 130 we subtract the -10.

130 - 10 = 120, which is our final answer.

$$
\begin{array}{r|r}
15 \diagdown +5 \\
\times\ 8 \diagup -2 \\
\hline
13 & -10 \\
\times 10 \\
\hline
130 & -10 = 120
\end{array}
$$

Let's take another set of numbers to understand this better.

Say, we have 17 x 9.

Step 1

17 and 9 are above and below the base of 10 respectively. 17 is greater than 10 by 7, so we write +7 and 9 is below 10 by 1, so we write -1.

$$
\begin{array}{c|c}
17 & +7 \\
\times\ 9 & -1 \\
\hline
 & \\
\end{array}
$$

Step 2

Now we cross-add or subtract, as the case may be. So we do either 17 - 1 = 16 or 9 + 7 = 16, whichever comes easily to you. So 16 becomes the first part of our answer.

We then multiply vertically on the right-hand side +7 and -1, giving us -7.

$$
\begin{array}{c|c}
17 & +7 \\
\times\ 9 & -1 \\
\hline
16 & -7 \\
\end{array}
$$

Step 3

To arrive at the final answer, we multiply the left-hand side by 10 (the base). So the left-hand side becomes 16 x 10 = 160. Now from 160 we subtract 7, giving us 153, our final and complete answer.

$$
\begin{array}{c|c}
17 & +7 \\
\times\ 9 & -1 \\
\hline
16 & -7 \\
\end{array}
$$

16 x 10 = 160
160 - 7 = 153

Let's take an example of a sum near the base 100. 102 x 99, for instance.

Step 1

We lay the sum down just as before. We also write the excess and deficiency from the base of 100.

$$
\begin{array}{r|l}
102 & +\,02 \\
\times\ 99 & -\,01 \\
\hline
\end{array}
$$

Step 2

We now cross-subtract. So we have 102 - 01 = 101 or we can also do 99 + 02 = 101.

We also multiply vertically on the right-hand side. So +02 x -01 = -02. Our sum now looks like this:

$$
\begin{array}{r|l}
102 & +\,02 \\
\times\ 99 & -\,01 \\
\hline
101 & -02 \\
\end{array}
$$

Step 3

Here our base is 100. So we multiply 101 by 100 (our base) giving us 10100. From it we subtract 02. So we have 10100 - 02 = 10098, our final and complete answer.

$$
\begin{array}{r|l}
102 & +\,02 \\
\times\ 99 & -\,01 \\
\hline
101 & -02 \\
\end{array}
$$

101 x 100 = 10100
10100 - 02 = 10098

Now let's take an example nearer to the base of 1000.

Say we have 1026 x 998.

Step 1
Let's lay the sum out just like before and write down the excesses and deficiencies.

$$
\begin{array}{r|r}
1026 & +026 \\
\times\ 998 & -002 \\
\hline
\end{array}
$$

Step 2
We cross-subtract and arrive at the left-hand side part of the answer. And then we multiply vertically on the right-hand side to get the right-hand side of the answer.

$$
\begin{array}{r|r}
1026 & +026 \\
\times\ 998 & -002 \\
\hline
1024 & -052 \\
\end{array}
$$

Step 3
We multiply 1024 by the base 1000 giving us 1024000. From it we take away -52, giving us 1023948, our final answer.

$$
\begin{array}{r|r}
1026 & +026 \\
\times\ 998 & -002 \\
\hline
1024 & -052 \\
\end{array}
$$

1024 x 1000 = 1024000
1024000 - 52 = 1023948

So till now we have developed an understanding of numbers below the base, above the base, and above and below the base. Let's continue exploring the concept deeper.

Multiples and Sub-Multiples

I was at IIT Madras giving a talk at their technical fest, Shaastra, and someone asked me this question: if two numbers are closer to 50 or 200 or any other multiple of 10, can we still apply the base method?

What do you think the answer could be? Think for a moment or two!

My answer to that question is 'yes!' We can still apply the base method, but with a minor adjustment.

Let's take an example and see how. Say we have to multiply 44 and 48, two numbers near to the base 50.

Step 1

We lay out the sum just as before. And write the excesses and deficiencies from our working base 50. So here we have 44, which is less than 50 by 6, and 48, which is less than 50 by 2.

$$
\begin{array}{r|r}
44 & -6 \\
\times\ 48 & -2 \\
\hline
&
\end{array}
$$

Step 2

We now cross-add or subtract as the case may be. So we have our left-hand side as 44 - 2 = 42 or 48 - 6 = 42. We then multiply vertically -6 and -2 to get 12. Our sum now looks like this:

$$
\begin{array}{r|r}
44 \diagdown & -6 \\
\times\ 48 \diagup & -2 \\
\hline
42 & 12
\end{array}
$$

Step 3

Now since our working base is 50, we have to adjust the sum for this. Since we get 50 by multiplying 10 and 5, we multiply 42 by 5 to get 210, which is the left-hand side of the answer.

Since our base is 50, which has one zero, the right-hand side should have one digit. And since there is one digit extra, we add that to the left-hand side.

210 + 1 = 211 and 2 comes down, giving us the answer of 2112.

$$44 \times -6$$
$$x\ 48 \quad -2$$

$$42 \quad | \quad 12$$

$$42 \times 5 = 210$$

2112

Let's see some more illustrations of the same principle at work. Say we have 49 x 47.

Step 1
Our working base here is 50 as both 49 and 47 are near to it. Let's lay the sum out and apply the rules learnt in the previous example. We also write down the excesses or deficiencies as before.

$$49 \quad | \quad -1$$
$$x\ 47 \quad | \quad -3$$
$$46 \quad | \quad 3$$

Step 2
Now since the working base is 10 x 5 = 50, we multiply the ten's place by 5. So here we multiply 46 (in the ten's place) by 5. This gives us 230.

$$49 \quad | \quad -1$$
$$x\ 47 \quad | \quad -3$$
$$46 \quad | \quad 3$$
$$x5$$
$$230$$

Step 3

Our final step will just be bringing the 3 down and combining it with 230 to make 2303.

Our final answer is 2303.

$$
\begin{array}{r|r}
49 & -1 \\
\times\ 47 & -3 \\
\hline
46 & 3 \\
\times 5 & \\
\hline
230\ 3
\end{array}
$$

Let me give you three worked-out sums as illustrations.

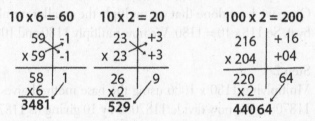

10 x 6 = 60

$$
\begin{array}{r|r}
59 & -1 \\
\times 59 & -1 \\
\hline
58 & 1 \\
\times 6 & \\
\hline
3481
\end{array}
$$

10 x 2 = 20

$$
\begin{array}{r|r}
23 & +3 \\
\times 23 & +3 \\
\hline
26 & 9 \\
\times 2 & \\
\hline
529
\end{array}
$$

100 x 2 = 200

$$
\begin{array}{r|r}
216 & +16 \\
\times 204 & +04 \\
\hline
220 & 64 \\
\times 2 & \\
\hline
440\ 64
\end{array}
$$

Now that we have understood the concept of the working bases let's move on to the final part of this base method. What happens when we have two different bases as part of the problem?

For example, say we have 981 x 93. How do we work this out?

Step 1

Find out the ratio of the bigger base to the smaller base. In this sum the base of 981 is 1000 and the base of 93 is 100. So dividing the bigger base by the smaller base we get 10. We multiply 93 by 10, giving us 930.

We now solve the sum 981 x 930 in the usual base method way.

Step 2
981 x 930 gives us 912330.
Because we multiplied 930 by 10, we divide the answer now by 10.
So 91233 is our answer figure.

Let's take another sum, this time 1006 x 118.

Step 1
We find out the ratio of the bigger base to the smaller base. Once we have done that we multiply the smaller number by it. So 118 x 10 = 1180. We now multiply 1180 and 1006.

Step 2
Multiplying 1180 x 1006 using the base method gives us 1187080. We now divide 1187080 by 10 giving us 118708 as our final answer.

We can now move on to a more general approach to multiplication. This is a very important maths sutra—**vertically and crosswise**. This, I think, is the king of all sutras, because of its myriad applications. With this maths sutra you can multiply or square numbers and find out the unknowns in a simultaneous equation. That's the reason I call it the king of all sutras!

Using the vertically and crosswise sutra you can multiply any number by any number, mentally or in one line. You can do this even left to right or right to left—the direction does not matter and you always get the correct answer if you apply the steps correctly.

So here we go. Let's see how to multiply two-digit figures by two-digit figures.

Two-Digit-by-Two-Digit Multiplication

In this type of multiplication we follow a visual pattern of multiplication, shown by the dots given here. Let's understand it thoroughly.

Say we have to multiply 12 x 43.

Step 1
We multiply vertically first, as shown by the dots in the illustration.

$$\begin{array}{r} 12 \\ \times\ 43 \\ \hline 6 \end{array}$$

We multiply 3 x 2 = 6. We get 6 in the unit's place.

Step 2
We will now multiply crosswise and add the sums as shown in the figure below.

So here we add (3 x 1) to (4 x 2) = 3 + 8 = 11. We put down 1 in the ten's place and carry 1 to the next step.

$$
\begin{array}{r}
12 \\
\times\ 43 \\
\hline
{}_{1}16
\end{array}
$$

Step 3

In this final step, we multiply vertically again, but this time we multiply $4 \times 1 = 4$. We now add the carry-over 1 with 4, giving us 5 as our hundred's place. Our complete answer is 516. Simple?

$$
\begin{array}{r}
12 \\
\times\ 43 \\
\hline
5{,}16
\end{array}
$$

Let's explore the pattern deeper.

Step 1

We multiply the units by the unit's place to arrive at the unit's place digit.

Step 2

We now need the ten's digit. So, to arrive at that, we multiply units by tens and tens by units as shown in the figure below.

Step 3

To finally arrive at the hundred's place we multiply the ten's place by the ten's place vertically as shown in the figure here.

2-Digits Vertically & Crosswise Pattern

Let us take another illustration. Say we have 78 x 69.

Step 1

As shown in the dot diagram here, we first multiply vertically in the unit's place. Nine times 8 equals 72. We keep the 2 in the unit's place and carry-over 7 to the ten's place.

$$
\begin{array}{r}
78 \\
\times\ 69 \\
\hline
_{7}2
\end{array}
$$

Step 2

In our second step, we multiply crosswise as shown here. We do (9 x 7) plus (6 x 8) together. This gives us 63 + 48 = 111. To 111 we add 7, our carry-over from the previous step, making it 118.

So, as before, we put 8 down in the ten's place and carry 11 over to the final step.

$$
\begin{array}{r}
78 \\
\times\ 69 \\
\hline
_{11}8_{7}2
\end{array}
$$

Step 3

In our final step, we multiply 6 x 7 giving us 42. We then add 11, our carry-over from the previous step. 42 + 11 = 53. We put 53 down and our final answer 5382 is right before us.

$$
\begin{array}{r}
78 \\
\times\ 69 \\
\hline
53_{11}8_{7}2
\end{array}
$$

So now we have understood how to multiply any two-digit number by another two-digit number with the help of the maths sutra vertically and crosswise method. The question now arises what does one do if one has to multiply any two-digit number by a single-digit figure?

This is nothing but child's play with the help of our maths sutra, vertically and crosswise.

Two-Digit-by-One-Digit Multiplication

Say we have to multiply 34 and 8. Let's take the sum and lay it out just as before.

Step 1
In our first step, we put a zero before 8. We do this because we want 8 to be a two-digit figure, so that it fits our requirement and we can apply the vertically and crosswise sutra.
So the sum first looks like this:

$$
\begin{array}{r}
34 \\
\times\ 08 \\
\hline
\end{array}
$$

Step 2
We apply the vertically and crosswise sutra now. First multiplying 8 and 4 together, we get 32. We put 2 in the unit's place and carry over 3 to the next step.

$$
\begin{array}{r}
34 \\
\times\ 08 \\
\hline
3^2
\end{array}
$$

In the next step, we simply do the crosswise step. So we have (8 x 3) + (0 x 4) = 24. To this we add the previous carry-over, 3. 24 + 3 = 27. We put 7 in the ten's place and carry over 2 to the next step.

Our sum looks like this now:

$$\begin{array}{r} 34 \\ \times\ 08 \\ \hline {}_2 7_3 2 \end{array}$$

Step 3

Vertically is our final step. Here we multiply 0 x 3 = 0. We then add 2 from the previous step to 0, giving us 2. Our answer becomes 272.

$$\begin{array}{r} 34 \\ \times\ 08 \\ \hline 2_2 7_3 2 \end{array}$$

Now that we have seen the vertically and crosswise for two digits, it's time we learnt the three-digit-by-three-digit multiplication too.

Three-Digit-by-Three-Digit Multiplication

Let's be a little attentive here in understanding the technique of the three-digit by three-digit multiplication pattern. This is new and might take a couple of minutes for anyone to understand. The good news is that we already know four of the five steps, as done earlier. So we just have to focus on understanding the pattern inbuilt here.

See the illustration. As always the dots represent the numbers over here.

3-Digit Vertically & Crosswise Pattern

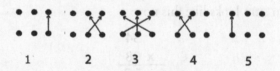

1   2   3   4   5

1) The first step is 'vertically'. As shown in the illustration, we multiply vertically the unit's digit by the other unit's digit to arrive at the unit's digit of our answer.

2) In our second step, we multiply crosswise to arrive at the ten's place digit.

3) This is a new step which we can call the 'star' step, simply because the pattern looks like a star. In this step, we have to arrive at the hundred's place. So we multiply and add as shown here to get to the hundred's place.

4) In our fourth step, we multiply crosswise as shown.

5) In our final step, we multiply vertically to arrive at the final answer figures.

Now let's apply the pattern to a three-digit-by-three-digit problem. Say for example 123 x 456, which is my personal favourite.

Step 1

Our first step in the sum 123 x 456 is vertically. We multiply 6 and 3 giving us 18. We put 8 down in the unit's place and carry over 1 to the next step.

$$
\begin{array}{r}
123 \\
\times\ 456 \\
\hline
{}_{1}8
\end{array}
$$

Step 2

In our second step, we multiply crosswise $(6 \times 2) + (5 \times 3)$. This gives us $12 + 15 = 27$. We also add the carry-over 1 from the previous step. So we have $27 + 1 = 28$. We put 8 down and carry-over 2 to the next step.

$$\begin{array}{r} 123 \\ \times\ 456 \\ \hline {}_2 8{}_1 8 \end{array}$$

Step 3

In the third step, we do the 'star' multiplication. Try and follow the steps here. We have:

$$(6 \times 1) + (4 \times 3) + (5 \times 2) = 6 + 12 + 10 = 28$$

To 28 we add the carry-over 2 from the previous step. So $28 + 2 = 30$. We put down 0 in the hundred's place and carry-over 3 to the next step.

$$\begin{array}{r} 123 \\ \times\ 456 \\ \hline {}_3 0{}_2 8{}_1 8 \end{array}$$

Step 4

The fourth and the fifth steps are the easiest. We just do crosswise here. So we have:

$$(5 \times 1) + (4 \times 2) = 5 + 8 = 13$$

To 13 we add the carry-over from the previous step, which is 3.
$13 + 3 = 16$. We put 6 down and carry over 1 to the next final step.

$$\times \quad \begin{matrix} 123 \\ \times\ 456 \\ \hline {}_16{}_30{}_28{}_18 \end{matrix}$$

Step 5

In our last and final step, we multiply vertically. So here
we have $4 \times 1 = 4 + 1$ (carry-over) = 5.

5 is the final digit of the answer which we bring down.
Our complete answer is 56088.

$$\begin{matrix} 123 \\ \times\ 456 \\ \hline 5{}_16{}_30{}_28{}_18 \end{matrix}$$

Let's take another three-digit-by-three-digit sum and see how
it works out using the vertically and crosswise sutra.

Say we have 231 x 745.

Step 1

We multiply vertically the units by the unit's place, just as
before, to arrive at the unit's answer digit. So $5 \times 1 = 5$. We
put down 5 in the unit's place.

$$\begin{matrix} 231 \\ \times\ 745 \\ \hline 5 \end{matrix}$$

Step 2

We do the crosswise step now to arrive at the ten's place
digit. So now we do $(5 \times 3) + (4 \times 1) = 15 + 4 = 19$. We
put down 9 in the ten's place and carry over 1 to the next
step. Remember, to get to the ten's place digit, we multiply

all the unit's places by the digit in the ten's place.

$$\begin{array}{r} 231 \\ \times\ 745 \\ \hline {}_{1}95 \end{array}$$

Step 3

In the third step, we do the 'star' multiplication. Try and follow the steps here. We have:

$(5 \times 2) + (7 \times 1) + (4 \times 3) = 10 + 7 + 12 = 29 + 1$ (carry-over) $= 30$.

We put down 0 in the hundred's place and carry over the 3 to the next step.

$$\begin{array}{r} 231 \\ \times\ 745 \\ \hline {}_{3}095 \end{array}$$

Step 4

We know the fourth step or crosswise step well enough now to arrive at the thousand's place digit.

So we have $(4 \times 2) + (7 \times 3) = 8 + 21 = 29 + 3$ (carry-over) $= 32$.

We put 2 down and carry over 3 to the next step.

$$\begin{array}{r} 231 \\ \times\ 745 \\ \hline {}_{3}2{}_{3}0{}_{1}95 \end{array}$$

Step 5

In our final step, we multiply vertically 7 and 2 to get 14. To 14 we add the carry-over 3 from the previous step making

it 17, which we put down as our final answer digit.
So the answer becomes 172095.

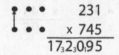

$$
\begin{array}{r}
231 \\
\times\ 745 \\
\hline
17{,}2{,}0{,}9\,5
\end{array}
$$

I know that with every sum you practise, you will get better and better. You have to give the new maths sutra methods some time to sink in. You have had years of practice in the traditional way, so you may be good at it and may be making mistakes in this method. Give this method some time and I promise you will be able to do it in one line.

Remember, as Vince Lombardi said, 'Quitters never win and winners never quit.'

Three-Digit-by-Two-Digit Multiplication

Now that we know how to multiply a three-digit number by a three-digit number, let us see an example of how we can multiply a three-digit number by a two-digit number.

All we have to do is put a zero in front of the two-digit number, thereby making it a three-digit number and then apply the vertically and crosswise method to it.

We now take an example 321 x 42.

Step 1
We first put a zero in front of 42, making it a three-digit number 042, and then we apply the vertically and crosswise sutra. We multiply 2 by 1 to get 2 in the unit's place of our answer.

$$
\begin{array}{r}
321 \\
\times\ 042 \\
\hline
2
\end{array}
$$

Step 2
We know the second step very well—crosswise. Here we quickly multiply $(2 \times 2) + (4 \times 1) = 4 + 4 = 8$. We put down 8 as our answer digit in the ten's place.

$$
\begin{array}{r}
321 \\
\times\ 042 \\
\hline
82
\end{array}
$$

Step 3
The third step is the star step! We do $(2 \times 3) + (0 \times 1) + (4 \times 2) = 14$. We put 4 in the hundred's place and carry over 1 to the next step.

$$\begin{array}{r} 321 \\ \times\ 042 \\ \hline {}_{1}482 \end{array}$$

Step 4

We just do crosswise here to arrive at the thousand's place digit.

$(4 \times 3) + (0 \times 2) = 12 + 1$ (carry-over) $= 13$. We put down 3 in the thousand's place and carry over 1 to the next step.

$$\begin{array}{r} 321 \\ \times\ 042 \\ \hline {}_{1}3{,}482 \end{array}$$

Step 5

This is where we do our vertical step multiplying 0 by 3, which gives 0. We add the carry-over 1 to 0 giving us 1 as the final answer figure.

Our answer is 13482.

$$\begin{array}{r} 321 \\ \times\ 042 \\ \hline 1{,}3{,}482 \end{array}$$

I hope that you enjoyed this chapter on multiplication the Vedic way!

Positive Affirmations in Mathematics

There is a lot of negative talk at school and in the environment around us, which disturbs our self-confidence and our ability to do things. Some of these remarks go like this:

- You are good for nothing.
- You are a fool.
- You can't do this sum right.
- You are a failure.
- You will always get bad grades in school and disappoint us.

These negative words affect our day-to-day life and they play like a video over and over again in our mind. It is therefore important to build our self-confidence in maths through something called Positive Affirmation Exercises.

A positive affirmation basically means talking positive to ourselves. This not only boosts our morale and confidence but it also ensures that we achieve positive results in our academics.

Try saying these words to yourself:

- I am a genius in maths.
- I have got 100 per cent marks in maths.
- I have a natural ability to learn maths.

At this point you may say, 'Sir, I get only 50 per cent in maths. How can I lie to myself?'

Well, firstly, you are not lying to yourself. It is a message to your brain that you believe you are good at something. Just thinking that makes you feel better and hence, your grades improve.

For example, tell yourself that you are the richest man in the world. Now how does that make you feel? Good, right?

Imagine what you could do with all the money in the world!

In the same way, positive affirmations make you feel good and help you achieve your results if you maintain a positive frame of mind.

Repeat these words five times each.
- I have done the sums correctly and have got 100 per cent in maths.
- I am doing maths sums in my mind.
- I am a genius in mathematics and am a maths guru.
- I am the best (feel good about it).
- All my friends respect me for my mathematical skills.

Now try to take a sum, say for example, 98 x 97, which you just learnt.

Try to close your eyes and think about the method by which you could solve this sum. See if you can get the correct answer by solving it in your mind.

The correct answer is 9506.

Addition

ON 15 MARCH 2015, an article about the importance of mental addition appeared in *India Today,* which shook the entire nation: 'Kanpur girl calls off marriage after groom fails ridiculously simple maths test.'

A girl in Kanpur walked out of her wedding when the groom failed to solve a simple maths problem she had posed. She had asked him to add 38 to 23. When he replied 51, she called off the marriage!

Well that's the importance of mental addition for you!

We all know how to do right-to-left addition. It was one of the first things we learnt as kids. The maths sutras will teach us how to add mentally and, that too, from left to right. Well you may ask why we need to do it from left to right. Firstly, solving the sum left to right makes the solution of the sum easier, especially in your head. Sometimes in competitive exams we need to know the left-hand side digit first. This is where the concepts of the maths sutras related to addition come handy.

So without any delay, let's go on and see how it's done by taking a sum like 78 + 45. We lay the sum down the normal way:

$$
\begin{array}{r}
78 \\
+\ 45 \\
\hline
\end{array}
$$

Step 1
We start adding left to right. 7 + 4 = 11. We keep this figure in our head!

$$
\begin{array}{r}
78 \\
+\ 45 \\
\hline
11
\end{array}
$$

Step 2
We now add 8 and 5 which equals 13. Remember that these steps are absolutely mental. Here you say to yourself, '11 and 13'.

Our sum now looks like this:

$$
\begin{array}{r}
78 \\
+\ 45 \\
\hline
11,13
\end{array}
$$

Step 3
In our final step, we 'combine' or add the middle digits
of the two figures as shown by the arrow. This step is
completely mental. So our answer becomes 123.

$$
\begin{array}{r}
78 \\
+\ 45 \\
\hline
11,13 \\
\hline
\mathbf{123}
\end{array}
$$

Simple enough?

I experienced a breakthrough moment in Durban, South
Africa, with respect to teaching this to fifth graders in a pilot
project with the Government of South Africa. The kids there
were struggling with basic addition. So we did little actions to
make them understand and remember the method like moving
hands and saying aloud the word 'combine'. The exercise bore
fruit! I realized that if we can combine hand actions with the
sutra, children tend to have a better recall and hence, achieve
better results.

Let's take another sum, say 87 + 69. We lay out the sum like this:

$$87$$
$$+ \ 69$$

Step 1
We add mentally 8 + 6 = 14. We hold the figure 14 in our mind.

$$87$$
$$+ \ 69$$
$$\overline{14}$$

Step 2
Now we add 7 + 9 = 16.

$$87$$
$$+ \ 69$$
$$\overline{14,16}$$

Step 3

In the final step we will add or 'combine' the middle digits
4 and 1 in our minds. So our answer becomes 156.

$$\begin{array}{r} 87 \\ + 69 \\ \hline 14,16 \\ \hline 156 \end{array}$$

So I will try to summarize what you need to say in your mind.
You will say, '14 and 16—combine, combine—156.' I hope
this is crystal clear.

Let's take another sum 48 + 97. Okay, so now try and do it
mentally just as shown here.

Let your mental voice say '13 and 15—combine—145
answer.' It's that simple.

But I think I will still show the steps—this time a little
quickly.

$$\begin{array}{r} 48 \\ + 97 \\ \hline \end{array}$$

$$\begin{array}{r} 48 \\ + 97 \\ \hline 13 \end{array}$$

$$\begin{array}{r} 48 \\ + 97 \\ \hline 13,15 \end{array}$$

$$\begin{array}{r} 48 \\ + 97 \\ \hline 13,15 \\ \hline 145 \end{array}$$

That was easy!

So far we have been doing two-digit additions. Let's take the case of three-digit figures now.

Left-to-Right Addition

The method which we just learnt can be applied to three-digit and four-digit problems as well. Let's take an example and see how. Say we have 582 + 759.

$$\begin{array}{r} 582 \\ + 759 \\ \hline \end{array}$$

Step 1

We start solving the sum by adding the digits in the first column. We add $5 + 7 = 12$ and we keep the figure in our mind. Our sum now looks like this.

$$\begin{array}{r} 582 \\ + 759 \\ \hline 12 \end{array}$$

Step 2

We now add the digits in the middle column. We add $8 + 5 = 13$. We now have two figures in our mind — 12, 13. We will now combine and add the middle digits, just like in the case of two-digit additions. So we have 133 in our mind now, after adding the middle digits.

$$\begin{array}{r} 582 \\ + 759 \\ \hline 12,13 \\ 133 \end{array}$$

Step 3

In our final step, we add 2 + 9 in the final column. We get 11. We will combine 133 and 11. The answer will be right in front of us as **1341.**

```
   582
 + 759
 ─────
 12,13
 133,11
 1341
```

This method is new and can be done mentally as I showed here. With a little bit of practice, you will be okay. Let's take another example and understand it better. Say we have 983 + 694.

```
   983
 + 694
```

Step 1

Here we add the first column digits from left to right. 9 + 6 = 15. We keep this figure 15 in our mind.

```
   983
 + 694
 ─────
   15
```

Step 2

Now we add the digits of the middle column, 8 + 9 = 17. We combine 15, 17 giving us 167. Our sum now looks like this:

```
   983
 + 694
 ─────
 15,17
  167
```

Step 3
We add the last column, 3 + 4 = 7. We already have 167 in
our mind and we bring down 7 making the answer 1677.
There is no need to combine here as 7 is a single digit. So
our answer is 1677.

$$
\begin{array}{r}
983 \\
+\,694 \\
\hline
15{,}17 \\
167{,}7 \\
1677
\end{array}
$$

Let's now do some difficult addition sums using this method.

Rapid Left-to-Right Columnar Addition

Say, for example, we have 5273 + 7372 + 6371 + 9782, we lay
out the sum as follows:

$$
\begin{array}{r}
5273 \\
7372 \\
6371 \\
+\,9782 \\
\hline
\end{array}
$$

Step 1
This sum is to be done exactly the same way we have learnt
in this chapter. We move column by column from left to
right. Here we add 5 + 7 + 6 + 9 = 27. We keep 27 in our
mind and move on to the next column.

```
     5273
     7372
     6371
   + 9782
     27,
```

Step 2

We add the second column $2 + 3 + 3 + 7 = 15$.

We now combine 27 from the first column to 15 from this column. This gives us 285. Our sum looks like this now:

```
     5273
     7372
     6371
   + 9782
     27͜15
     285
```

Step 3

We will now move on to totalling the third column. $7 + 7 + 7 + 8$ gives us 29.

In our mind, we already have 285. So we now combine 285, 29. So this gives us 2879. Our sum now looks like this:

```
     5273
     7372
     6371
   + 9782
     27͜15
    285͜29
    2879
```

Step 4

In our final step, we total up the final column. We get 3 + 2 + 1 + 2 = 8. Now because 8 is a single digit, we simply just bring it down. So our answer becomes 28798. Simple?

$$
\begin{array}{r}
5273 \\
7372 \\
6371 \\
+\ 9782 \\
\hline
27{,}15 \\
285{,}29 \\
\hline
28798
\end{array}
$$

Let's see another sum of a similar nature.

Let's take the following numbers 8336 + 4283 + 3428 + 9373 and start adding them.

$$
\begin{array}{r}
8336 \\
4283 \\
3428 \\
+\ 9373 \\
\hline
\end{array}
$$

Step 1

Just like before we add 8 + 4 + 3 + 9 = 24 from the first column. We keep this in our mind.

$$
\begin{array}{r}
8336 \\
4283 \\
3428 \\
+\ 9373 \\
\hline
24
\end{array}
$$

Step 2

We add the second column from the left, 3 + 2 + 4 + 3 = 12. We now combine and add. So we have 24, 12. We get

252, which we keep in our mind for a while.

```
    8336
    4283
    3428
  + 9373
   24,12
    252
```

Step 3

We take the third column and add the digits in it, $3 + 8 + 2 + 7 = 20$. We combine this with the numbers we got. So we have 252, 20. We combine and we get 2540 like this:

```
    8336
    4283
    3428
  + 9373
   24,12
   252,20
    2540
```

Step 4

In our final step, we add the digits in the right-most column, $6 + 3 + 8 + 3 = 20$. We get 20 and then we combine it with 2540. So we have 2540, 20 giving us 25420 as our answer. The sum looks like this:

```
    8336
    4283
    3428
  + 9373
   24,12
   252,20
  2540,20
   25420
```

In my experience, I have noticed that people find the numbers just below a multiple of 10 quite difficult to add mentally! But actually, numbers that are just below a multiple of 10 are very easy to add if you know the relevant maths sutra. I am referring to numbers like 9, 19, 18, 48, 57, etc.

Say, for example, we have 24 + 9.

Solving this is actually quite easy if you apply the maths sutra 'By Addition and by subtraction'.

See 9 is very close to 10 and they differ by 1 only. So if we add 10 to 24, which is quite easy to do, and then subtract 1 from the total, which is also very simple to do, we can then arrive at our answer. The sum looks like this:

$24 + 9 = 24 + 10 = 34 - 1 = 33$, our answer.

Let's take another sum. Say we have 56 + 8.

We apply the same rules here as well. 8 is very close to 10. We can add 10 to 56 and take away 2 to arrive at our answer figure. So we have:

$56 + 10 = 66 - 2 = 64$, our answer.

I hope it is clear as to why we take away 2, after adding 10.

Let's take a different set of addition problems, based on the same principles.

Say we have 46 + 18.

Here we add 20 to 46 making it 66 and then subtracting 2, giving us 64 as our answer figure. Our sum looks like this:

$46 + 18 = 46 + 20 = 66 - 2 = 64$.

Another illustration would be 168 + 19.

We apply the same maths sutra here, by addition and by subtraction.

So we have 168 + 19 = 168 + 20 = 188 - 1 = 187, our answer.

Supposing we have 458 + 38.

We apply the same maths sutra, by addition and by subtraction.

So we have 458 + 40 = 498 - 2 = 496, our answer.

I also wanted to point out that we could do this sum in other ways too. Say, like this:

458 + 38 = 460 + 40 = 500 - 2 - 2 = 496, our answer.

That's the choice Vedic maths gives you. There can be more than one way to solving a sum. You choose the one that you like most and the one which comes most naturally to you.

The Importance of Goals

If we have to progress in life we must learn how to set goals. We get a sense of direction for our work with the setting of goals and we know where we stand and where can we go from there.

So in this short section, we will learn how to set goals, after which we will see how we can improve our maths score.

Suppose you are currently getting about 60 per cent as your maths score. This means you have mastered 60 per cent of the paper, which implies that those topics are your strengths. Also the sums you cannot do or the sums that are incorrect fall under your weaknesses.

The trick here is to convert your weaknesses into strengths. This method is not known to many students, but this very concept is an opportunity to change your life from here on and never looking back.

Step one is to make a list of the maths topics that you know thoroughly.

Say, for example,

- Multiplication
- Addition
- Subtraction

Step two is to make a list of maths topics that form your weaknesses.

Say, for example,

- Division
- Linear Equations
- Simultaneous Equations

Now your goal is to convert all these topics into your strengths.

For example, here is a list:

Goal-Setting for Maths

Strengths	Weaknesses
Multiplication	Division
Addition	Linear Equations
Subtraction	Simultaneous Equations

Now you have to convert your weaknesses into your strengths.

Through goal-setting you can remove your weaknesses.

Weaknesses	Study time Devoted	Tests Attempted	Confidence	Strengths
Division	2 hours	5	Yes	Yes
Linear Equations	1.5 hours	0	No	No
Simultaneous Equations	2.5 hours	1	No	No

Make a table like this. And your goal in maths is to get 100 per cent. Now the only way you can get to 100 per cent is by converting your weaknesses into strengths.

In this table, you must write the devoted study time for each chapter. You must also write the number of tests done and the confidence level for that particular topic. So as you work on the table every day your weaknesses will be converted into strengths!

3

Subtraction

LET ME ASK YOU a question, how much is 123 x 999? Solve it using the base method!

Well, as you are attempting the sum, honestly tell me, did you have a problem subtracting 123 from 1000? I have asked this question in almost all my workshops and to diverse student groups as well as maths teachers, and the answer has always been a 'Yes!'

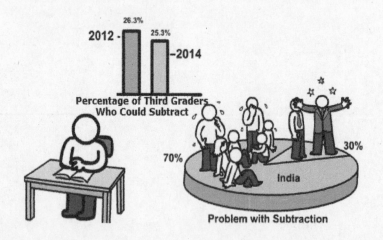

Percentage of Third Graders Who Could Subtract

2012 — 26.3%

25.3% —2014

Problem with Subtraction

70% 30% India

Let me validate this with a report. In India in 2012, 26.3 per cent of third graders could do a two-digit subtraction. This number is at 25.3 per cent in 2014 according to *ASER*. Just think about the struggle that over 70 per cent of 1.22 billion Indians face with subtraction.

In light of this, why can't we have a new and alternate subtraction method introduced to replace the current methods, since they have clearly failed? If we continue to do the same thing over and over again, we will get the same result, over and over again.

So why not make subtraction simple?

So, ladies and gentlemen, I present to you a new maths sutra for subtractions—All from 9 and last from 10. Let's learn its applications.

Subtractions Using All from 9 and Last from 10

Say we have to solve 1000 - 283.

Here we solve left to right and not right to left and we will apply the maths sutra for subtractions, all from 9 and last from 10. So let's begin our subtraction.

$$1\ 0\ 0\ 0\ -\ 2\quad 8\quad 3$$

$$\downarrow\qquad\downarrow\qquad\downarrow$$

Subtract   Subtract   Subtract
from 9   from 9   from 10

$$=\ 7\qquad 1\qquad 7$$

So our answer is 1000 - 283 = 717

Step 1

The maths sutra is all from 9 and last from 10, which means that when subtracting we will subtract all the numbers from 9 but the last one from 10.

So in the first step, we have 9 - 2 = 7. This is the first answer digit.

Step 2
In the second step, we again subtract 8 from 9. So we have 9 - 8 = 1. This is our second answer digit.

Step 3
In our last step, we will now subtract from 10. So we subtract 3 from 10. We have 10 - 3 = 7, which is our final answer digit.

So our answer is 717.

Let us take another example. Say we have 1000 - 476.

Step 1
We apply the maths sutra, all from 9 and last from 10, and arrive at the answer.

So here we have left to right, 9 - 4 = 5.

5 is the first answer digit from the left.

Step 2
Now in our second step, we will subtract 9 - 7 = 2, the middle digit of our answer.

Step 3
Here we end, so we will subtract from 10.

So we have 10 - 6 = 4.

Our final answer is 524.

$$1\ 0\ 0\ 0\ -\ 4\qquad 7\qquad 6$$

Subtract Subtract Subtract
from 9　from 9　from 10

$$=\ 5\qquad 2\qquad 4$$

So our answer is 1000 - 476 = 524

At this point, I would like to share something. This method will apply to all powers of ten, like 10, 100, 1000, 10000, 100000, etc. This is an easy way to subtract from base numbers like these.

So, for example, if we have 10000 - 1234, we will apply the same maths sutra once again. Let's see how.

Step 1
We start subtracting from left to right. So first we have 9 - 1 = 8.

Step 2
We then have, 9 - 2 = 7.

Step 3
9 - 3 = 6.

Step 4
Finally, the last term 4 will be taken away from 10.
So 10 - 4 = 6.
Our final answer is 8766.

Let's now take three quick examples and let us do it mentally this time!

1) 1000 - 547

Yes! I am sure you can do this sum mentally. Just apply the rule for all from 9 and last from 10, and our answer is 453, that's it!

2) 1000 - 346

We apply the same maths sutra just as before. So we get 654 as our answer.

3) 10000 - 5389

Here even though the numbers are four digits or more, we can still apply the same maths sutra. So our answer is 4611.

Just remember that this method is applied only when we are subtracting from powers of 10 like 10, 100, 1000, 10000, etc.

Having learnt the maths sutra all from 9 and last from 10, you must be wondering how to subtract any number from any number — mentally.

Super Subtraction

This is the time when I introduce you to Super Subtraction. You can subtract any number from any number mentally using this technique. One of my friends, Dr B.S. Kiran, attempted a Guinness record for subtracting a seventy-digit number from another seventy-digit number in sixty seconds flat! He used the same method that I am sharing with you now. For fun we can call it Super Subtraction!

Let's take an example first. Say we have 651 - 297. The sum must be written in the normal way only.

$$
\begin{array}{r}
651 \\
- 297 \\
\hline
\end{array}
$$

Step 1

We will subtract from left to right. We have 6 - 2 = 4.

$$
\begin{array}{r}
651 \\
- 297 \\
\hline
4
\end{array}
$$

Step 2

In our second step, we see that 9 at the bottom is higher than 5 on the top. So the result will be a negative number. What we do is reduce 4 from the previous step to 3 and then carry over 1 to the next step.

We prefix 5 with 1 and it becomes 15. 15 - 9 = 6, this is our middle digit.

Our sum looks like this:

$$\begin{array}{r} 6\overset{\text{\scriptsize 1}}{5}1 \\ -\ 297 \\ \hline \cancel{4} \\ 36 \end{array}$$

Step 3

In our final step and in the third column, we see that 7 at the bottom is greater than 1 on the top. We repeat the same thing we did in step 2.

Since there is a negative in the final column, we go back one column and reduce 6 to 5 and then carry over 1, prefixing it to 1. This makes it 11.

11 - 7 = 4. So our answer becomes 354.

$$\begin{array}{r} 6\overset{\text{\scriptsize 1}}{5}\overset{\text{\scriptsize 1}}{1} \\ -\ 297 \\ \hline \cancel{4} \\ 3\cancel{6}54 \end{array}$$

Essentially, we are doing the same steps involved in traditional subtraction, but with a little difference. You can just call out the answer digits by this maths sutra and go about this in a speedy way!

Let's take another sum and solve it with this new maths sutra. Say we have 425 - 168.

$$\begin{array}{r} 425 \\ -168 \\ \hline \end{array}$$

Step 1
We will subtract from left to right. So we have 4 - 1 = 3, we put down 3 as shown here.

$$
\begin{array}{r}
425 \\
-168 \\
\hline
3
\end{array}
$$

Step 2
In the next step, we see that the result will be a negative number. 6 below is greater than 2 above. So we go back one step, reduce 3 to 2 and carry over the 1 to 2. We prefix 1 to 2, making it 12. 12 - 6 = 6. We put down 6 as the middle digit of our answer. Our sum now looks like this:

$$
\begin{array}{r}
4\overset{1}{2}5 \\
-168 \\
\hline
326
\end{array}
$$

Step 3
In our final step, again we see a negative happening as 8 below is greater than 5 above. So we make the middle digit 6 a 5 and carry over 1 to 5 making it 15.

15 - 8 = 7.

Our answer becomes 257.

$$
\begin{array}{r}
4\overset{1}{2}\overset{1}{5} \\
-168 \\
\hline
32657
\end{array}
$$

Oral Subtraction

You can actually call out the answer digits left to right orally.
Let's see how for the sum of 425 - 168.

Step 1: 4 - 1 = 3

Step 2: 2 - 6 = negative, so 3 becomes 2, and then we do
12 - 6 = 6

Step 3: 5 - 8 = negative, so 6 becomes 5, and the last figure
becomes 15 - 8 = 7

The final answer is 257.

Let's take another sum, say this time, a four-digit subtraction
sum 7643 - 4869.

$$\begin{array}{r} 7643 \\ - 4869 \\ \hline \end{array}$$

Step 1
We start subtracting from left to right.
7 - 4 = 3

$$\begin{array}{r} 7643 \\ - 4869 \\ \hline 3 \end{array}$$

Step 2
In the second step, we see that 6 - 8 will result in a negative
number.
So we just reduce 3 to 2 from the previous step and carry
over 1 to 6, making it 16.

Then 16 - 8 = 8.
So far our answer is 28.

$$7643$$
$$- 4869$$
$$328$$

Step 3

In our third step, we see that 6 at the bottom is greater than 4 on the top. So we go back one step and make 8 a 7 and prefix 1 to 4 on the top. 14 - 6 = 8.

So our answer figures so far are 278.

$$7643$$
$$- 4869$$
$$32878$$

Step 4

In our final step, we see that 9 at the bottom is more than 3 on top.

Therefore, we go back one step and make 8 a 7 and carry over 1, prefixing it to 3.

13 - 9 = 4, this is our final answer digit.
So our complete answer is 2774.

$$7643$$
$$- 4869$$
$$3287874$$
$$2774$$

I hope you have all understood this maths sutra for subtraction. Let me now share with you a slightly bigger sum that you can do with the help of this sutra. I have purposely taken a bigger sum, because I just wanted to show you how much subtraction power you have using this method. If you can do this, then two-digit or three-digit subtractions become a piece of cake.

Let's solve 638475 - 429763

$$
\begin{array}{r}
638475 \\
-429763 \\
\hline
\end{array}
$$

Step 1
We do normal subtraction from left to right.
6 - 4 = 2 and 3 - 2 = 1.
We put 2 and 1 as our answer digits like this:

$$
\begin{array}{r}
638475 \\
-429763 \\
\hline
21
\end{array}
$$

Step 2
In the third column from the left, we see that 9 below is greater than 8 above, so we do what we did earlier. We make 21 into 20 and carry over 1, prefixing it to 8 to make it 18.
18 - 9 = 9
So we have 209 in our head now.

$$
\begin{array}{r}
6\overset{1}{3}8475 \\
-429763 \\
\hline
2\overset{}{1}09
\end{array}
$$

Step 3

Again in the fourth column, we see that 7 below is greater than 4 above.

So we go back one place and reduce 209 to 208.

We carry over 1, prefixing it to 4. It becomes 14. 14 - 7 = 7.

We keep a mental note of 2087.

$$\begin{array}{r} 638\overset{1}{4}75 \\ -429763 \\ \hline 2\overset{}{1}0987 \end{array}$$

Step 4

Next we do simple subtraction.

7 - 6 = 1 and 5 - 3 = 2.

Our complete answer is 208712.

$$\begin{array}{r} 638\overset{1}{4}75 \\ -429763 \\ \hline 2\overset{}{1}098712 \end{array}$$

I think now you are ready to do subtractions that will keep us in good stead when we learn the maths sutra for division. There is also an opportunity for you to make a record of sorts using super subtraction.

If you can subtract a seventy-digit number from a seventy-digit number in less than a minute, contact the Guinness World Records and get your feat recorded there! Until then practise, my fellows, for practice and practice alone makes a man perfect!

Now let's practise some instant subtractions mentally! I am sure you can do them with your eyes closed, because the concept is so very easy.

So what is 82 - 8?

Remember this concept? 82 - 10 = 72 + 2 = 74. Yes, that's our answer!

Let's take another sum, say 94 - 17. Do this sum with your eyes shut and with the Vedic maths approach.

94 - 20 = 74 + 3 = 77

If you got this answer mentally—then you're awesome. If not, try and go through this chapter again and I am sure you will get it.

Listening and Maths Skills

Do you know that in mathematics 90 per cent of your success depends on how good a listener you are? If you listen properly to your maths teacher teaching, chances are that you will do well, but if you are lost in thoughts about the movie you saw last night, then chances are that your maths scores won't be so good. So, listening is important for you to be good in mathematics.

Like, for example, if you are just hearing me now and not listening to me, chances are that you haven't understood the topic we did right now.

I give you five points to be a better listener.

1) In your maths class, try to sit up with your back straight and be attentive.

2) Maintain eye contact with the teacher. Don't lose interest easily!

3) Convey your interest in class by genuinely nodding.
4) If you don't understand any topic feel free to ask questions. Some students feel shy to ask questions, because they feel their friends will laugh at them. Asking questions is a sign of intelligence. If you don't ask questions you will never know what's right or wrong.
5) Do not interrupt the teacher! Wait for the teacher to finish and then respectfully ask your questions. And then you will soon start getting good grades.

Cracking maths in competitive exams is easy if your fundamentals are in place. If your fundamentals are not in place, prepare them with this series of Vedic maths classes.

4

Division

Today, as I write, this problem has stumped the Internet and has gone viral too. Cheryl's birthday challenge was meant to test the better high-school students competing in the Singapore and Asian Schools Olympiad.

I reproduce the problem here so that you too can attempt it.

Albert and Bernard just met Cheryl. 'When's your birthday?' Albert asked Cheryl. Cheryl thought for a second and said, 'I'm not going to tell you, but I'll give you some clues.' She wrote down ten dates:

15 May, 16 May, 19 May,

17 June, 18 June,

14 July, 16 July,

14 August, 15 August, 17 August.

'My birthday is one of these,' she said. Then Cheryl whispered in Albert's ear the month—and only the month—of her birthday. To Bernard, she whispered the day—and only the day.

'Can you figure it out now?' she asked Albert.

Albert replied, 'I don't know when your birthday is, but I know Bernard doesn't know, either.'

Bernard added, 'I didn't know originally, but now I do.'

Albert said, 'Well, now I know, too!'

When is Cheryl's birthday?

But enough about Cheryl for now, let us now dive deep into the concept of division by the maths sutras.

In division, the maths sutra used is called the Flag Method. We are going to learn this method of division in two parts. In the first part, we will learn the steps that are important and in the second part, we will learn how to solve any division problem.

Division by the Flag Method

Say we have to divide 848 by 31. Here, 848 is our dividend and 31 is our divisor. This is what the layout of the sum will be:

$$3^1 | 8 \ 4 \ 8$$

Here, if you notice, in our divisor 31, we have written 1 on top raised up on the flag. This is not 3 to the power of 1, but 3 raised to 1. 3 is our flagpole and 1 is our flag.

And since we have one digit in our flag, we go one digit to our left and put a stroke. The stroke signifies where the decimal point will be. Starting the division process, we will have to remember two rules.

Rule 1: We divide by the flagpole.

Rule 2: We subtract the sum of the flag multiplied by the previous quotient digit , and subtract it from the result of Rule 1.

Now let's go back to our problem and try to solve it. We have
848 divided by 31.

Step 1

We divide 8 by 3. It gives us 2 as the quotient digit and 2 as
the remainder. We put the quotient digit in its appropriate
place. Then we take the remainder 2 and prefix it to 4,
making it 24.

$$3^1 \mid 8\,_24 \mid 8$$
$$\overline{ 2}$$

Step 2

In our next step, we will subtract the sum of the flag
multiplied by the previous quotient digit from 24.
The flag is 1 and the previous quotient digit is 2. (1 x 2 =
2). So 24 - 2 = 22.

Step 3

Now that one cycle of division and subtraction is over, we
will repeat the cycle and divide again. So we divide 22 by
3, giving us 7 as the quotient digit and 1 as the remainder.
This remainder 1 we prefix to 8, making it 18.

$$3^1 \mid 8\,_24 \mid {}_18$$
$$\overline{ 2 \quad 7}$$

Step 4

Continuing the steps, we will now subtract (1 x 7 = 7)
from 18.
18 - 7 = 11. And 11 divided by 3 is 3 with remainder 2.

We put down 3 and take the remainder 2 to the next step.
Our answer unfolds before us, 27.3.

$$3^1 \begin{array}{c|c|c} 8_24 & 18_2 \\ \hline 27 & 3 \end{array}$$

Our answer is 27.3
to 1 decimal place.

Although this method may seem difficult, it will give
awesome results if practised properly and also save time during
exams. It will even help solve problems in data interpretation.
This method gives you the answer in one line without the trial
and error that exists in the traditional method.

You just need to divide and subtract. These are the two
steps that are vital for you to get the correct answers using this
method.

Let's take another sum. Say we have to divide 651 by 31.

We will lay the sum out as before and start solving it.

$$3^1 \begin{array}{c|c|c} 6 & 5 & 1 \\ & & \end{array}$$

Step 1
We divide 6 by the flagpole 3. This gives us 2 as the quotient
and 0 as the remainder. We prefix 0 to 5 making it 05.

$$3^1 \begin{array}{c|c|c} 6_05 & 1 \\ \hline 2 & \end{array}$$

Step 2

In our second step, we will subtract.

So we have $5 - (1 \times 2) = 5 - 2 = 3$.

We make a mental note of 3.

Step 3

We now divide.

3 divided by 3 is 1, which we put down as our next quotient digit, and remainder is 0, which we prefix to 1.

$$\begin{array}{c|cc|c} 3^1 & 6_0 5 & 0 1 \\ \hline & 2 \ 1 . & 0 \end{array}$$

Step 4

We will now subtract $(01 - (1 \times 1)) = 0$.

0 divided by 3 is 0.

So our answer is 21.0 exactly.

$$\begin{array}{c|cc|c} 3^1 & 6_0 5 & 0 1 \\ \hline & 2 \ 1 . & 0 \end{array}$$

Let's now take a new sum $5576 \div 25$.

We lay out the sum just like before.

$$\begin{array}{c|ccc} 2^5 & 5 \ 5 \ 7 \ 6 \\ \hline & \end{array}$$

Step 1

We start dividing now.

5 divided by 2 gives us 2 as the quotient and the remainder as 1.

We prefix the remainder 1 to 5, making it 15.

$$2^5 \overline{\left| 5\,{}_15\ 7\ \right|6}$$
$$ 2 $$

Step 2

After division, comes subtraction.

So from 15 we subtract (5 x 2) = 15 - 10 = 5.

We make a mental note of 5.

Step 3

After subtraction, we now divide.

5 divided by 2 is 2 and the remainder is 1. We put 2 as the new quotient digit and prefix 1 to 7, making it 17.

$$2^5 \overline{\left| 5\,{}_15\,{}_17\ \right|6}$$
$$ 2\ 2 $$

Step 4

Now we will subtract.

17 - (5 x 2) = 17 - 10 = 7

After subtraction, we will now divide by the flagpole.

So we have 7 divided by 2, giving us 3 and remainder 1, which we prefix to 6 to make it 16.

$$2^5 \overline{\left| 5\,{}_15\,{}_17\,{}_16 \right.}$$
$$ 2\ 2\ 3 $$

Step 5

We will now subtract after division.

16 - (5 x 3) = 16 - 15 = 1

We will now divide.

So we have 1 divided by 2, giving us 0 as the quotient and
1 as the remainder, which we take over to the next step. To
carry on the sum, we will put a 0 and make it 10, like this:

$$2^5 \quad | \quad 5\ _15\ _17\ |_16\ _10$$
$$\overline{\ 2\ 2\ 3\ \cdot\ |\ 0}$$

Step 6
In this step, we will first subtract.
$10 - (5 \times 0) = 10 - 0 = 10$
Dividing 10 by 2 we get 5 as the next quotient digit.
Our answer is 223.05.

$$2^5 \quad | \quad 5\ _15\ _17\ |_16\ _10$$
$$\overline{\ 2\ 2\ 3\ \cdot\ |\ 05}$$

If you are still having difficulty getting this, it's because
you haven't understood the steps. If you just focus on the
steps you will get it 100 per cent. Let me explain it to you
one more time.

Rule 1: We divide by the flagpole.
Rule 2: We subtract the sum of the flag multiplied by the
previous quotient digit.

Let's see another sum $2924 \div 72$.
Try and do this yourself for practice.

$$7^2 \quad | \quad 2\ 9\ 2\ |\ 4$$

Step 1

The first step is division.

So we divide 29 by 7. It gives us a quotient digit of 4 and remainder 1, which we prefix to 2, making it 12.

$$7^2 \overline{\left| \begin{array}{ccc} 2\ 9 & {}_1 2 & 4 \\ \hline 4 & & \end{array} \right.}$$

Step 2

We will now subtract. We subtract the sum of the flag multiplied by the previous quotient digit.

So we have $12 - (2 \times 4) = 12 - 8 = 4$.

4 divided by 7 gives us 0 as the quotient digit and 4 as the remainder. We prefix 4 to 4, making it 44.

$$7^2 \overline{\left| \begin{array}{ccc} 2\ 9 & {}_1 2 & {}_4 4 \\ \hline 4\ 0 & {}\cdot & \end{array} \right.}$$

Step 3

Our next step is to subtract.

$44 - (2 \times 0) = 44 - 0 = 44$

We will now divide by the flagpole 7. So 44 divided by 7 is 6 with remainder 2.

To carry on the sum, we will put a 0 and make it 20, like this:

$$7^2 \overline{\left| \begin{array}{ccc} 2\ 9 & {}_1 2 & {}_4 4 {}_2 0 \\ \hline 4\ 0 & {}\cdot & 6 \end{array} \right.}$$

Step 4

We will now subtract.

$20 - (2 \times 6) = 20 - 12 = 8$

8 divided by 7 is 1, which we put down. So our answer, correct to two decimal places, is 40.61.

$$7^2 \;\Big|\; 2 \;\; 9 \;_{\!1}2 \;\Big|_4 4_2 0$$
$$ 4 \;\; 0 \;\cdot\; 6 \;\; 1$$

Now that we have got the hang of these steps, let us now see a concept of Altered Remainders, which will help us in solving problems involving negative values.

Altered Remainders

Let's take a sum say $43 \div 8$.

Here our quotient is 5 and the remainder is 3.

But suppose we want to increase our remainder, how can we do that?

Well, we can do so by decreasing the quotient. So it will look like this:

Quotient	Remainder
5	3
4	$3 + 8 = 11$
3	$11 + 8 = 19$
2	$19 + 8 = 27$
1	$27 + 8 = 35$

Here we have successively decreased our quotient and simultaneously increased our remainder.

Let's see another example. Say we have $28 \div 5$.

Suppose now we want to increase our remainder, how can we do that?

Again we do so by simply decreasing the quotient.

Quotient	Remainder
5	3
4	3 + 5 = 8
3	8 + 5 = 13
2	13 + 5 = 18
1	18 + 5 = 23

So we see that decreasing the quotient will increase the remainder.

We now move on to another section of this division method, where we will use the concept of altered remainders to our benefit.

Let's take a sum. Say we have $3412 \div 24$.

Before I solve it, I would like you to give it a shot. See if you get a negative value in your calculations. And that's where I step in.

We lay the sum out the usual way and start solving it. Since the flag has one digit (4) we move one digit to our left and put the decimal stroke before 2 and after 1.

$$2^4 \mid 3 \; 4 \; 1 \mid 2$$

Step 1
We first divide 3 by 2. It gives us 1 as the quotient and 1 as

the remainder, which we prefix to 4 to make it 14.

$$2^4 \mid \begin{array}{c} 3\,_14\ 1 \end{array} \mid 2$$
$$\overline{\ 1}$$

Step 2
After division, we will now subtract.
$14 - (4 \times 1) = 14 - 4 = 10$
We make a mental note of this.

Step 3
After subtraction, we will now divide.
We have 10 divided by 2, giving us 5 as the quotient and 0 as the remainder, which we prefix to 1, making it 01.

$$2^4 \mid \begin{array}{c} 3\,_14\,_01 \end{array} \mid 2$$
$$\overline{\ 1\ 5}$$

Step 4
Now after division, we are going to subtract.
So we have $01 - (4 \times 5) = 01 - 20 = -19$.
Now we have a negative value in our sum and hence, we can't go on with the usual steps.
So we go back one step and alter our remainder by decreasing the quotient.
We decrease 5 to 4 and increase the remainder 0 to 2.
This makes it $21 - (4 \times 4) = 21 - 16 = 5$.
We make a mental note of 5.

$$2^4 \mid \begin{array}{c} 3\,_14\,_21 \end{array} \mid 2$$
$$\overline{\ 1\ \cancel{5}\,4}$$

Step 5

So we will now divide.

5 divided by 2 gives us 2 as the next quotient digit and the remainder is 1, which we prefix to 2 to make it 12.

$$2^4 \ \left|\ 3\,_14\,_21\ \right|\,_112$$
$$\overline{\qquad 1\;5\,4\,2\cdot\qquad}$$

Step 6

We will now subtract $12 - (4 \times 2) = 12 - 8 = 4$.
We keep 4 in our mind.

Step 7

4 divided by 2 is 2, which is our next quotient digit and the remainder is 0, which we prefix to 0.

$$2^4 \ \left|\ 3\,_14\,_21\ \right|\,_112\,_00$$
$$\overline{\qquad 1\;5\,4\,2\cdot 2\qquad}$$

Step 8

Now we have a negative value. Note that the sum doesn't end here because of the 00.

In fact, the next step is $00 - (4 \times 2) =$ negative.

So we go back one place and alter the quotient digit 2 to 1 and that gives us the answer of 142.1.

$$2^4 \ \left|\ 3\,_14\,_21\ \right|\,_112\,_00$$
$$\overline{\qquad 1\;5\,4\,2\cdot \ |21\qquad}$$

Step 9

We can move on and continue adding zeroes to get to the other decimal digits following the same steps of division and subtraction.

Let's take another illustration 5614 ÷ 21.

We lay the sum out just like before. Based on the flag, we keep the decimal stroke at the appropriate place — before 4 and after 1. Now this is what our sum looks like:

$$2^1 \mid 5\ 6\ 1 \mid 4$$

Step 1
We start dividing by 2.
So we have 5 divided by 2, giving us the quotient digit of 2 and remainder 1, which we prefix to 6 to make it 16.

$$2^1 \mid 5_{\ 1}6\ 1 \mid 4$$
$$2$$

Step 2
After division, we will now subtract. So we have:
16 - (1 x 2) = 16 - 2 = 14
We keep 14 in our mind.

Step 3
After subtraction, we will now divide.
So we have 14 divided by 2, which gives me 7 as the next quotient digit.
We put 7 down at the appropriate place and prefix 0 to 1, making it 01.

$$2^1 \mid 5_{\ 1}6_{\ 0}1 \mid 4$$
$$2\ 7$$

Step 4

In our next step, we have 01 - (1 x 7) = negative.

Since we have a negative value, we alter our quotient digit from 7 to 6 and change the remainder digit from 0 to 2, making it 21.

The sum now looks like this:

$$2^1 \,\Big|\, 5\,_16\,_21 \,\Big|\, 4$$
$$\overline{2\,\!/\!7\,6}$$

Step 5

We will now subtract.

21 - (1 x 6) = 21 - 6 = 15

We keep 15 in our mind.

Step 6

And now we divide.

We have 15 divided by 2, giving us 7 as the next quotient digit.

We have a remainder of 1, which we prefix to 4 to make it 14.

$$2^1 \,\Big|\, 5\,_16\,_21 \,\Big|\, _14$$
$$\overline{2\,\!/\!7\,67.}$$

Step 7

Now we will subtract.

14 - (1 x 7) = 14 - 7 = 7

We keep 7 in our mind.

Step 8

And after subtraction, we must divide again.

So we divide 7 by 2, giving us 3 as the next quotient digit and 1 as the remainder.

Our answer becomes 267.3.

$$\begin{array}{c|c|c} 2^1 & 5{,}6{,}1 & 4{,}0 \\ \hline & 2\!\!\!/67. & 3 \end{array}$$

Now let us take a new illustration with altered remainders.

$7943 \div 42$

We lay the sum out as before. Since there is one digit in the flag, we will put our decimal stroke before 3 and after 4 in our dividend.

$$\begin{array}{c|c|c} 4^2 & 7 \ 9 \ 4 & 3 \\ \hline & & \end{array}$$

Step 1
We will start by dividing by 4.

So 7 divided by 4 gives the quotient digit of 1 and the remainder 3, which we prefix to 9 to make it 39.

Our sum looks like this now:

$$\begin{array}{c|c|c} 4^2 & 7 \ _39 \ 4 & 3 \\ \hline & 1 & \end{array}$$

Step 2
Now we will subtract.

So we have $39 - (1 \times 2) = 37$.

We keep 37 in our mind.

Step 3
Now we will divide.

37 divided by 4 equals 9 with remainder 1, which we prefix to 4 to make it 14.

$$\begin{array}{c|c|c} 4^2 & 7 \ _39 \ _14 & 3 \\ \hline & 1 \ 9 & \end{array}$$

Step 4

Now we subtract. So we have 14 - (9 x 2) = negative.

We apply the concept of the altered remainder because of the negative.

So we go back one step and change 9 to 8 and increase the remainder by 4 making it 5.

$$4^2 \,|\, 7\,_3 9\,_5 4\,|\, 3$$
$$\overline{1\,9\!\!/\,8\,|\quad}$$

Step 5

So now we have 54 - 16 = 38.

38 divided by 4 gives 9 with remainder 2, which we prefix to 3 making it 23.

Our sum looks like this now:

$$4^2 \,|\, 7\,_3 9\,_5 4\,|_2\, 3$$
$$\overline{1\,9\!\!/\,8\,9\,|\quad}$$

Step 6

We will now subtract.

So we have 23 - (2 x 9) = 23 - 18 = 5.

We keep 5 in our mind.

Step 7

In our final step, we will divide.

So we have 5 divided by 4, giving us 1 as both the new quotient digit and remainder.

So our answer becomes 189.1.

$$4^2 \,|\, 7\,_3 9\,_5 4\,|_2\, 3\,_1 0$$
$$\overline{1\,9\!\!/\,8\,9\,.\,|1}$$

Auxiliary Fractions

We use auxiliary fractions to further simplify our division processes. 'Auxiliary' means providing supplementary or additional help and support. This section will also help us in calculating fractions to exact decimal places in a short time, helping to save time during examinations.

We shall now see how to solve fractions whose denominators end in 9.

When the denominator of the fraction ends in 9

For the fraction $\frac{6}{49}$, note that 49 is closest to 50. So the auxiliary fraction is arrived at by dividing the numerator 6 by 50. This gives us the auxiliary fraction of $\frac{0.6}{5}$.

Step 1

We now divide 0.6 by 5, which gives us the quotient 0.1 and remainder 1.

This is how we write it:

$$\frac{6}{49} = 0_1.1$$

Step 2

After writing this down, we take the next dividend as 1.1 and divide by 5. Now we get the next digit 2 and remainder 1. This is how we write it:

$$\frac{6}{49} = 0_1.1_12$$

Remember that the remainder is written just before 2.

Step 3
We now take 12 as the dividend and divide it by 5. This again gives us 2 as the quotient and the remainder is also 2, which we write like this:

$$\frac{6}{49} = 0_1.1_12_22$$

We carry on this operation until we get the required number of decimal places.

There's one more step.

Step 4
We now divide 22 by 5, and it gives us the quotient 4 and remainder 2.

Our sum now looks like this:

$$\frac{6}{49} = 0_1.1_12_22_24.....$$

We now carry on this operation until we get the required number of decimal places.

Our answer is 0.1224, rounded off.

Let's take another example, say, $\frac{11}{149}$.

Here we note that 149 is close to 150. So 149 is replaced by 15 and 11 is replaced by 1.1.

So we have $\frac{11}{149} = AF \frac{1.1}{15}$

Step 1

We now divide 1.1 by 15. We get the quotient 0 and remainder 11. We put this 11 just before 0 and it becomes 110.

Our sum now looks like this:

$$0._{11}0$$

Step 2

We now divide 110 by 15. We get the quotient 7 and remainder 5. We prefix this remainder 5, making the next dividend 57:

$$0._{11}0_57$$

Step 3

We continue the same process. We now divide 57 by 15. We get the quotient 3 and remainder 12.

We prefix the remainder 12 to 3. Our sum looks like this:

$$0._{11}0_57_{12}3$$

We can now continue this until we have the required number of decimal places.

Step 4

So our last dividend is 123 which we again divide by 15. We get the quotient 8 and remainder 3.

Finally, our sum looks like this:

$$0._{11}0_57_{12}3_38$$

Our answer is 0.0738.

Let's take yet another sum. Say we have $\frac{16}{19}$.

We now convert $\frac{16}{19}$ into an auxiliary fraction. Here we note that 19 is close to 20. 19 is replaced by 2 and 16 is replaced by 1.6. Our auxiliary fraction will be:

$$\frac{16}{19} = AF \; \frac{1.6}{2}$$

Step 1

We divide 1.6 by 2. We get the quotient 8 and remainder 0, which we put just before 8. Our sum now looks like this:

$$0._08$$

Step 2

We divide 08 by 2. This gives us the quotient digit 4 and remainder 0 again.

So we write down 4 and prefix 0 to it.

Our sum looks like this now:

$$\frac{16}{19} = 0._08_04$$

Step 3

We now divide 04 by 2. This gives us the quotient digit of 2 and remainder 0 again. We write down 0 and prefix 0 to 2 like this:

$$\frac{16}{19} = 0._08_04_02$$

Step 4

We now divide 02 by 2 which gives the quotient 1 and remainder 0 which we prefix to 1.

Our sum now looks like this:

$$\frac{16}{19} = 0._08_04_02_01....$$

So our exact answer is 0.8421.

5

Digit Sums

IN ANY COMPETITIVE examination, the accuracy and precision of your answers are the keys to success and a high percentile. Vedic mathematics gives you just that! In this chapter, we shall see how to check our answers whether it is multiplication, addition or subtraction. With the concept of digit sums, we can check our answers fairly quickly.

So What Is a Digit Sum?

The word 'digit' refers to numbers like 1, 2, 3, 4, 5, etc. And the word 'sum' means 'to add'. So combining the two, we have 'digit sum', which is nothing but the sum of the digits.

If we have to find the digit sum of, say 61, we will add 6 and 1, which gives us 7. So 7 is the digit sum of 61.

Say we have to find the digit sum of 92.

We add $9 + 2 = 11$. We add again $1 + 1 = 2$. So 2 is the digit sum of 92.

Let's take 568 for example. We add $5 + 6 + 8 = 19 = 1 + 9 = 10 = 1 + 0 = 1$. So 1 is the digit sum of 568.

Let's see the digit sums of a few more numbers.

Number	Summing Digits	Digit Sum
65	$6 + 5 = 11 = 1 + 1 = 2$	2
721	$7 + 2 + 1 = 10 = 1 + 0 = 1$	1
3210	$3 + 2 + 1 + 0 = 6$	6
67754	$6 + 7 + 7 + 5 + 4 = 29 = 2 + 9 = 11$ $= 1 + 1 = 2$	2
82571	$8 + 2 + 5 + 7 + 1 = 23 = 2 + 3 = 5$	5
1890	$1 + 8 + 9 + 0 = 18 = 1 + 8 = 9$	9
23477	$2 + 3 + 4 + 7 + 7 = 32 = 3 + 2 = 5$	5

This, as you can see, is a very simple and easy concept to understand. All you have to do is keep adding all the digits till you get a single-digit number and that will be the digit sum. The digit sum of a number helps us to check the answer, as we are about to see.

Casting Out Nines

The other method to find out the digit sum of any given number is called Casting Out Nines. In this method, we simply cast out 9 and the digits adding up to 9. Whatever remains, we add them up and that gives us the digit sum of the number.

For example:
1. We have to find the digit sum of 8154912320.
We will cancel 8 and 1, because they add up to 9.
We will also cancel 5 and 4, because they add up to 9 again. We will then, of course, cancel the digit 9.
So our sum looks like this now: 8~~154~~9~~12320.

We then add 1 + 2 + 3 + 2 + 0 = 8.

So the digit sum becomes 8.

2. Let's take another number 970230612.
We will cancel 9, 7, 2, 3 and 6.
Our sum now looks like this: 9~~70230612.
We will now add 1 + 2 = 3, which is our digit sum.

Using Digit Sums to Check Answers

Now that we have understood the concept of digit sums, we can move ahead and see how we can check our maths problems with this system.

Let's start with addition, taking 734 + 352.

The sum of these two numbers is 1086; let us check whether this is the correct answer or not.

$$734 \longrightarrow \overset{\text{Digit Sum}}{5}$$
$$+ \ 352 \longrightarrow +1$$
$$\overline{1086} \longrightarrow \overline{6}$$

Step 1

We first find out the digit sum of 734, which is $7 + 3 + 4$ $= 14 = 1 + 4 = 5$.

Now we will find the digit sum of $352 = 3 + 5 + 2 = 10 = 1 + 0 = 1$.

Adding these two, we get $5 + 1 = 6$.

Step 2

To find the digit sum of 1086, $1 + 0 + 8 + 6 = 15 = 1 + 5 = 6$

Now since both the digit sums match, we can say that our answer 1086 is correct.

Let us take another example, $2344 + 6235$.

The sum of these two numbers is 8579. Now let us check if this answer is correct or not using the digit sum method.

Step 1

We first find the digit sum of 2344, which is $2 + 3 + 4 + 4$ $= 13 = 1 + 3 = 4$.

Then we find the digit sum of 6235, which is $6 + 2 + 3 + 5 = 16 = 1 + 6 = 7$.

We will add both these digit sums. We get $4 + 7 = 11 = 1 + 1 = 2$.

Step 2

We will now find the digit sum of 8579, which is $8 + 5 + 7 + 9 = 29 = 2 + 9 = 11 = 1 + 1 = 2$.

Now since both the digit sums match, we can say that our answer 8579 is correct.

$$
\begin{array}{r}
& \text{Digit Sum} \\
2344 \longrightarrow & 4 \\
+6235 \longrightarrow & +7 \\
\hline
8579 \longrightarrow & 2
\end{array}
$$

Checking Subtractions

Let's take 4321 - 1786.

Our answer is 2535, but let's check this with the digit sum method.

$$
\begin{array}{r}
& \text{Digit Sum} \\
4321 \longrightarrow & 1 \\
- 1786 \longrightarrow & -4 \\
\hline
2535 \longrightarrow & -3
\end{array}
$$

$$-3 = 9 - 3 = 6$$

Step 1

The digit sum of 4321 is $4 + 3 + 2 + 1 = 10 = 1 + 0 = 1$.

The digit sum of 1786 is $1 + 7 + 8 + 6 = 22 = 2 + 2 = 4$.

We now do 1 minus 4, which gives us -3 $(1 - 4 = -3)$.

Now this becomes $9 - 3 = 6$.

Remember, adding or subtracting 9 from any number does not change the digit sum.

Because this is a negative digit sum, we add 9 and get 6 as our final digit sum.

Step 2
We now check with the answer 2535 and find the digit sum of it.
The digit sum of 2535 is $2 + 5 + 3 + 5 = 15 = 1 + 5 = 6$.

So we get 6 from both the calculations, which means our answer is correct!

Let us take another subtraction sum for checking 74637 - 24267.

$$
\begin{array}{r}
& \text{Digit Sum} \\
74637 \longrightarrow & 9 \\
-24267 \longrightarrow & -\ 3 \\
\hline
50370 \longrightarrow & 6 \\
\end{array}
$$

Our answer for this is 50370. We will now check this with the digit sum concept.

Step 1
The digit sum of 74637 is 9 and the digit sum of 24267 is 3.
We simply subtract both the digit sums and get $9 - 3 = 6$.

Step 2
The digit sum of 50370 is $5 + 3 + 7 = 15 = 1 + 5 = 6$.
Since both our digit sums match, we can safely say that our answer is correct!

Let us now move to the multiplication check in the same way.

Checking Multiplications

Let's take a sum, say 62 x 83.

Our answer is 5146. Now let's find the digit sum and check our answer.

$$
\begin{array}{ccc}
 & & \text{Digit Sum} \\
62 & \longrightarrow & 8 \\
\times 83 & \longrightarrow & \times 2 \\
\hline
5146 & \longrightarrow & 16 \\
& & = 1 + 6 = 7
\end{array}
$$

The digit sum of 62 is 8 and the digit sum of 83 is 2.
We multiply both the digit sums 8 x 2 and we get 16 which is 1 + 6 = 7.
The digit sum of 5146 = 16 = 1 + 6 = 7.
So our answer is correct.

Let's take another sum, say 726 x 471.

The product is 341946. Let us get it checked by digit sums.

$$
\begin{array}{ccc}
 & & \text{Digit Sum} \\
726 & \longrightarrow & 6 \\
\times 471 & \longrightarrow & \times 3 \\
\hline
341946 & \longrightarrow & 18 \\
& & = 1 + 8 = 9
\end{array}
$$

The digit sum of 726 is 6.
The digit sum of 471 Is 3.
6 x 3 = 18 = 1 + 8 = 9
The digit sum of our answer 341946 is also 9.
So our answer is correct.

A Word of Caution

Digit sum is an important checking tool, but it has its limitations.

For example, let's take 12 x 34.

Supposing we write our answer as 804. The digit sum of 804 is $8 + 0 + 4 = 12 = 1 + 2 = 3$.

But when we check our answer it gives us 408. Both 804 and 408 have the same digit sum of 3. This is a limitation of the digit sum method. Please take care to write the answer in the exact same order or else an error may occur.

$$
\begin{array}{ccc}
 & & \text{Digit Sum} \\
12 & \longrightarrow & 3 \\
\times 34 & \longrightarrow & \times 7 \\
\hline
804 & \longrightarrow & 21 \\
\downarrow & & = 2 + 1 = 3
\end{array}
$$

Wrong Answer

Habits and Maths

Anything that we do repeatedly becomes a habit! Practising maths, in particular, is something that should become a habit! Good habits give rise to positive thoughts. It also means good results, success and rewards in your school and college life.

I know many students who like to study at the last moment. They eventually come last. Maths toppers, especially those who crack the IITs and IIMs, practise maths regularly.

Let us take a look at some of the habits that prevent our success in the maths world.

- **Laziness and Procrastination**: Some of us leave everything for tomorrow, and that tomorrow sadly never comes. It is imperative to change that in order to be successful.

- **Accepting Failure as a Habit:** Some students just lose hope. They accept failure and give up. In some colleges, I have seen students rejoicing when they fail in groups. Your success and failure lie in your own hands. You are what you think. If you accept failure, it will come back to you. But if you challenge yourself and change your habits, the rewards and success will surely follow.

- **Blaming Parents and Teachers:** This habit is very common. We never take responsibility for our own mistakes and tend to blame someone else—our parents, our teachers, our friends, etc. We love the blame game.

Now let's see how positive habits shape our future.
- **Perseverance and Dedication:** If we regularly work towards attaining success in maths, we will achieve it sooner or later. There can be no substitute for hard work and perseverance. Make maths your habit and, trust me, success will come to you.

- **Collective Wisdom:** Whenever you discuss maths with your friends or teachers, you give birth to collective wisdom. This wisdom will help you scale newer heights in the field of mathematics.

- **Focus on Work:** Take it up and finish it! Have a 'do it now' attitude. This will help you to move on to higher topics and progressively scale new heights!

If you keep in mind these few nuggets of wisdom, then I guarantee a wonderful future for you. If you can make a habit of prioritizing which topics to do first and which a little later, I am sure all of life's challenges will be a cakewalk!

6

Fractions

'Success does not consist in never making mistakes, but in
never making the same one a second time.'
—George Bernard Shaw

CAT Aspirants

AND THAT'S ONE of the mantras to success in a limited-time competitive exam like the CAT or the GMAT. Never make the same mistake twice!

In this chapter, we are going to look into the maths sutra that will help in solving fractions.

Addition of Fractions

Type 1: With the same denominator

For the addition of fractions, whenever we have the same denominator, we simply add the numerators and simplify as much as possible.

For example, in this sum, we note that the denominators are the same, so we just add the numerators to arrive at our final answer.

$$\frac{5}{11} + \frac{3}{11}$$

So we have the sum as

$\frac{5 + 3}{11} = \frac{8}{11}$ and our answer is $\frac{8}{11}$.

Let us take another example. Say we have $\frac{3}{10} + \frac{1}{10}$.

Since the denominators are the same here as well, all we have to do is add up the numerators and then simplify.

Our sum looks like this: $\frac{3 + 1}{10} = \frac{4}{10} = \frac{2}{5}$

Type 2: *When one denominator is a factor of the other*

Let us take an example to understand this type of situation:

$$\frac{2}{5} + \frac{9}{20}$$

Here we see that 5 is a factor of 20. So we must convert $\frac{2}{5}$ into $\frac{8}{20}$ after multiplying the numerator and denominator by 4. Then, we will simply add the numerators. The sum looks like this:

$$\frac{2}{5} + \frac{9}{20}$$

$$= \frac{8}{20} + \frac{9}{20}$$

$$= \frac{17}{20}$$

Let us consider another example.

How about $\frac{11}{15} + \frac{7}{30}$?

Here we see that 15 is a factor of 30. So, we shall multiply the numerator and the denominator of $\frac{11}{15}$ by 2. Then, again, we will again just add the numerators. Our sum looks like this:

$$\frac{11}{15} + \frac{7}{30}$$

$$= \frac{22}{30} + \frac{7}{30}$$

$$= \frac{22 + 7}{30}$$

$$= \frac{29}{30}$$

Type 3: Using the vertically and crosswise method

Now we will learn how to add any two given fractions using the maths sutra vertically and crosswise.

Let's take any two fractions.

$$\frac{3}{5} + \frac{1}{4}$$

Step 1

We will multiply crosswise and add the results.

So this becomes $(4 \times 3) + (5 \times 1) = 12 + 5 = 17$. This is our numerator.

$$\frac{3}{5} \times \frac{1}{4}$$

Step 2

We will get our denominator by multiplying $5 \times 4 = 20$.

So our complete answer is $\frac{17}{20}$.

$$\frac{3}{5} \times \frac{1}{4}$$

Our sum looks like this:

$$\frac{3}{5} + \frac{1}{4}$$

$$= \frac{12 + 5}{20}$$

$$= \frac{17}{20}$$

Let's take another example, $\frac{2}{11} + \frac{7}{9}$.

Step 1

In our first step, we will multiply crosswise and add as shown here.

$$\frac{2}{11} \diagdown \frac{7}{9}$$

So this becomes $(9 \times 2) + (11 \times 7) = 18 + 77 = 95$.
95 is our numerator.

Step 2

We will get the denominator by multiplying 11 by 9. 11 × 9 = 99. So our denominator is 99.

$$\frac{2}{11} \diagdown \frac{7}{9}$$

Our sum looks like this:

$$\frac{(2 \times 9) + (11 \times 7)}{99} = \frac{95}{99}$$

So we could at least simplify fractions with the help of the maths sutra vertically and crosswise. This is easy and elegant, simple and sweet!

Let's now move on to subtraction of fractions.

Subtraction of Fractions

Type 1: With the same denominator

Let's start with $\frac{15}{16} - \frac{3}{16}$.

This is easy and simple and can be solved like this:

$$\frac{15-3}{16} = \frac{12}{16} = \frac{3}{4}$$

All we have to do here is subtract 15 - 3 = 12 and write the denominator as it is. Finally, we simplify $\frac{12}{16}$ to $\frac{3}{4}$.

Let's take another example, $\frac{13}{35} - \frac{6}{35}$.

First, we need to subtract 13 - 6 = 7. This is our numerator. Our denominator will remain 35.

So our answer becomes $\frac{13}{35} - \frac{6}{35} = \frac{7}{35} = \frac{1}{5}$.

Type 2: When one denominator is a factor of the other

Let's consider $\frac{1}{2} - \frac{5}{24}$.

If you look at the denominators, you will note that 2 is a factor of 24. And 2 x 12 = 24, so we multiply the top and the bottom of the fraction $\frac{1}{2}$ by 12.

So we get $\frac{12}{24} - \frac{5}{24}$, which becomes $\frac{12-5}{24} = \frac{7}{24}$.

Easy, isn't it?

Let's take another example, $\frac{7}{9} - \frac{2}{3}$.

3 x 3 = 9, so we multiply the numerator and denominator of $\frac{2}{3}$ by 3. We get $\frac{6}{9}$, and then we subtract normally.

So we get $\frac{7}{9} - \frac{2}{3} = \frac{7}{9} - \frac{6}{9} = \frac{1}{9}$ as our answer.

Type 3: Using the vertically and crosswise method

Like we just saw in the case of addition the vertically and crosswise method plays an important role in subtraction too.

Let's try it with $\frac{6}{7} - \frac{1}{2}$.

Step 1

We will first find the numerator. We do this by applying the rule crosswise. So we have (2 x 6) - (7 x 1) = 12 - 7 = 5. Our numerator is 5.

$$\frac{6}{7} \diagdown\!\!\!\!\diagup \frac{1}{2}$$

Step 2

We will now find the denominator. This will be $7 \times 2 = 14$.

Our answer will be $\frac{5}{14}$.

Let us take another sum, $\frac{12}{25} - \frac{3}{50}$.

Step 1

We will first find the numerator by multiplying crosswise and subtracting the results.

$$\frac{12}{25} \diagdown \frac{3}{50}$$

So we get $(12 \times 50) - (25 \times 3) = 600 - 75 = 525$. Our numerator is 525.

Step 2

Now, we will multiply the two denominators and add them. We get $25 \times 50 = 1250$ as our denominator.

$$\frac{12}{25} \diagdown \frac{3}{50}$$

Our fraction becomes $\frac{525}{1250}$, which we simplify to get $\frac{21}{50}$, which is our answer.

Seven Tips to Excel in Mathematics

How to be good at mathematics and get better grades?

I am sure you must have asked this question to yourself or someone else at some point in time.

So wait no more, because I am going to share with you seven tips to excel in mathematics!

1. Learn Vedic mathematics. This amazing system will help you build up your foundations in mathematics. You can do this by mastering the concepts shared in this book.
2. Do not try to learn everything in one go. Take a topic and slowly master it by practising problems on the topic.

3. Remember that practice makes perfect, and that nobody gets everything right all the time. If you mess up, keep reviewing.
4. Don't hesitate to ask teachers your doubts. Asking questions is a sign of intelligence. Everyone should ask questions.
5. Do not overwork and tire yourself out just before a maths exam. Make sure to get a good night's sleep and write your exam with a fresh mind.
6. Reward yourself when you get good scores in a maths exam.
7. If you have problems with basic maths concepts like multiplication tables, then make sure you memorize them, because a little bit of memorization does no harm.

7

Decimals

'A ship in port is safe, but that is not what ships are built for!'

—Admiral Grace Hopper

THERE IS SO much this lovely quote teaches us. This line applies to each one of us. Each one of us is like a ship. We have to face our problems and overcome them just as a ship faces waves in the open sea! Right now, our problem may be maths and tomorrow, it may be something else. We have to remember to keep on going and to keep achieving the best we can.

This chapter is about decimals and we will learn how to add, subtract, multiply and divide them in the quickest possible manner.

The Decimal System

In the decimal number system, the position of the number determines its value. If the number 4 is placed on the left of 7 to make it 47, this means four tens and not four ones. This is called the place value of a number.

Without place value, calculations would be extremely difficult. The principle of place value is that the value of the place immediately to the left of any given place is ten times as great. Also, a position to the right is ten times as small, or one-tenth of the value of the place immediately to the left.

We place the unit's column in the middle. The columns to the left of the units column are for tens, hundreds and so on, and the columns to the right are for the one-tenths, one-hundredths and so on.

Thousand	TH
Hundreds	H
Tens	T
Units	U
.	.
tenths	t
hundredths	h
thousandths	th

The Decimal Point

The decimal point is used to distinguish between whole

numbers and parts of a whole. For example, 0.1 is one-tenth of 1, 0.01 is one-hundredth of 1 and 0.001 is one-thousandth of 1. And so, 45.7 is four tens, five units and seven tenths.

Calculating with the Decimal Point

Calculations with decimals are done in the same way as with whole numbers. When reading the decimal numbers, we say 'seven point eight six' for 7.86. 8.12 is read out as 'eight point one two', 4.5 as 'four point five' and 0.04 as 'zero point zero four'.

Addition of Decimals

While adding or subtracting decimal numbers, keep the decimal points in a vertical line.

For example, 4.34 + 3.42

$$
\begin{array}{r}
4.34 \\
+3.42 \\
\hline
7.76
\end{array}
$$

You can do this sum from left to right or from right to left. You can also do this sum easily by first adding without the decimals and then putting the decimal point in the end, as shown here.

For 4. 34 + 3. 42, we first ignore the decimal point of the two numbers and add them as they are.

$$
\begin{array}{r}
434 \\
+342 \\
\hline
776
\end{array}
$$

We get 776 as our answer, in a sum which was fairly easy to do.

Now we put the decimal place in our answer. We note that in our numbers the decimal point is two places before the end digit. So, we put the decimal point two places before the end digit in the result. Our answer is 7.76.

Now say that we have to add 78.3 + 2.031 + 2.3245 + 9.2.

Note again that while solving the sum, we put the decimal points directly below one another, as shown here:

$$
\begin{array}{r}
78.3 \\
2.031 \\
2.3245 \\
+\ 9.2 \\
\hline
\\
\hline
\end{array}
$$

After adding, the decimal point will come in the same place. We add normally and get our answer as 91.8555.

$$
\begin{array}{r}
78.3 \\
2.031 \\
2.3245 \\
+\ 9.2 \\
\hline
91.8555
\end{array}
$$

Let's take another similar sum and see how it works.
Say we have 0.0004 + 6.32 + 1.008 + 3.452.

$$
\begin{array}{r}
0.0004 \\
6.32 \\
1.008 \\
+\,3.452 \\
\hline
10.7804 \\
\hline
\end{array}
$$

Here, we have put the decimal points one below the other, and then calculated by normal addition. Note the position of the decimal point. This is very important and the positioning of the decimal point will make all the difference.

Let's take another example, $5.004 + 0.302 + 20.489 + 1.07$.

$$
\begin{array}{r}
5.004 \\
0.302 \\
20.489 \\
+\,1.07 \\
\hline
26.865 \\
\hline
\end{array}
$$

Note the way decimals have been lined up one below the other. The positioning is vital when dealing with decimals. Our answer becomes 26.865 after normal addition.

Let's now move on to the subtraction of decimals.

Subtraction of Decimals

Just as it was with addition, while subtracting decimal numbers, we should always line up the decimal points one below the other. All the other columns will line up too—tens under tens, units under units and so on.

Let's take 45 - 2.09 and try to solve it.

$$45.00$$
$$- 2.09$$

Step 1
We first line up the decimal points one below the other as shown. 45 is written as 45.00, because 2.09 has two digits after the decimal points. So we pad with zeroes after the decimal point. The sum becomes 45.00 - 2.09.

Step 2
We subtract first without considering the decimal points. So we do 4500 - 209. This gives us 4291. We then put back the decimal point and our answer is 42.91.

$$45.00$$
$$- 2.09$$
$$42.91$$

Let's take another sum, 7.005 - 0.55.

Step 1
We line up the sum like this:

$$7.005$$
$$- 0.55$$

Note the positioning of the decimal points—they have to line up, one below the other.

Step 2

Now we pad with zeroes. We put a zero after 0.55. The advantage of putting zeroes is that we safely get our answer, because the units will be below the units, the tens under the tens and so on. So we have:

$$7.005$$
$$- 0.55\mathbf{0}$$

Note the 0 after 0.55. Our sum has become 7.005 - 0.550.

Step 3

We subtract first without considering the decimal points. So we do 7005 - 0550. This gives us 6455. We then put back the decimal point and our answer is 6.455.

$$7.005$$
$$- 0.550$$
$$\overline{6.455}$$

See how simple it is to subtract with decimals? Decimals are now child's play! Let's take another example and understand subtractions with decimals better.

Let's consider 19.19 - 3.3.

Step 1

We first line up the decimal points one below the other. Again, I would like to emphasize that this first step is the most important because of the positioning. We must ensure that the decimal points are one below the other. Otherwise, we may end up with an incorrect answer.

$$
\begin{array}{r}
19.19 \\
- \ \ 3.3 \\
\hline
\\
\hline
\end{array}
$$

Step 2
Now we pad with zeroes. We put a zero after 3 and before 3, as shown below. So we have:

$$
\begin{array}{r}
19.19 \\
- \ 03.30 \\
\hline
\\
\hline
\end{array}
$$

Step 3
We subtract first without considering the decimal points. So we perform 1919 - 0330. This gives us 1589. We then carefully put back the decimal point in the proper place and get 15.89 as our answer.

$$
\begin{array}{r}
19.19 \\
- \ 03.30 \\
\hline
15.89 \\
\hline
\end{array}
$$

In my experience, people have a decimal phobia. So, through the maths sutras, I have tried to make the topic as friendly as possible. One point to note here is that while performing addition and subtraction, we can use the Vedic methods that we have already learnt and then put in the decimal point. Thus, our fundamentals become stronger.

As we have learnt, decimals are all about positioning. After we line up the decimal points one below the other, we pad the numbers with zeroes. Then we perform our calculations to find the correct answer.

So let's move on to multiplying decimals.

Multiplication by 10, 100, 1000, Etc.

Multiplication by powers of ten is essentially very easy as it just involves moving the decimal point.

Let's multiply 7.86 by 10. Note that ten has one zero. So, just move the decimal point one place to the right to make it 78.6.

Now let's multiply 7.86 by 100. We see that hundred has two zeroes. So, we just move the decimal point two places to the right to get 786 as our answer.

Here are examples of what happens to a number when it's multiplied by different powers of ten.

Number	x 10	x 100	x 1000	x 10000
0.72	7.2	72	720	7200
0.91	9.1	91	910	9100
0.04	0.4	4	40	400
9.25	92.5	925	9250	92500
2.34	23.4	234	2340	23400
5.04	50.4	504	5040	50400
12.36	123.6	1236	12360	123600
42.03	420.3	4203	42030	420300
561.321	5613.21	56132.1	561321	5613210

Multiplication of Decimals

It's very simple to multiply any decimal by any decimal using the vertically and crosswise method, which we learnt in the chapter on multiplication.

So whenever we have to multiply any two-digit decimal by another two-digit decimal, we shall use the vertically and crosswise pattern and then do the sums.

Let's take 7.3 x 1.4.

$$
\begin{array}{r}
7.3 \\
\times\,1.4 \\
\hline
\end{array}
$$

Step 1

Ignoring the decimals at first, we start multiplying from right to left. We multiply 4 x 3 = 12. Put down 2 and carry 1.

$$
\begin{array}{r}
7.3 \\
\times\,1.4 \\
\hline
1\,2 \\
\end{array}
$$

Step 2

Cross multiply now (4 x 7) + (1 x 3) = 28 + 3 = 31 plus the carry-over 1 = 32. We put down the 2 and carry 3.

$$
\begin{array}{r}
7.3 \\
\times\,1.4 \\
\hline
3\,2\,2 \\
\end{array}
$$

Step 3

Next, we multiply vertically 1 x 7 = 7 and add the carry-over 3 to give us 10, which we put down to get 1022. Next, we have to put the decimal place. We see that in each of the numbers the decimal point is after one place from the right. So in the answer we place our point 2 places from the right.

$$
\begin{array}{r}
7.3 \\
\times\,1.4 \\
\hline
10.3_21\,2
\end{array}
$$

Our answer becomes 10.22.

NOTE: To put the decimal place in the correct place, always count the number of places from the right in both the numbers. Add them up, count the total number of decimal places in then product and then put the decimal point.

Let's take another example to get a better picture of the process. We'll take 6.2 x 5.4.

$$
\begin{array}{r}
6.2 \\
\times\,5.4 \\
\hline
 \\
\hline
\end{array}
$$

Step 1

We ignore the decimals in the beginning and apply the vertically and crosswise formula to our problem to get our answer. We multiply 2 x 4 to get 8. We put down 8 in the units place. Our sum looks like this:

$$6.2$$
$$\times 5.4$$
$$\overline{8}$$

Step 2

We now multiply crosswise $(4 \times 6) + (5 \times 2) = 24 + 10 = 34$. So we put down 4 in the ten's place and carry 3 over to the next step. Remember, so far we have been ignoring the decimals.

$$6.2$$
$$\times 5.4$$
$$\overline{_348}$$

Step 3

In the final step, we multiply vertically again. So we multiply $6 \times 5 = 30$. To this, we add the carry-over 3. So it becomes $30 + 3 = 33$. Our answer so far is 3348.

Now we have to place the decimal point. Any guesses as to where we shall put the decimal point?

$$6.2$$
$$\times 5.4$$
$$\overline{33_348}$$

We count the number of places from the right in both the numbers. Add them up to get the total number of decimal places we have to put in our answer. We see that in each of the numbers the decimal point is after one place from the right. So in the answer, we place our point two places from the right.

Our final answer is 33.48.

I hope this is clear till here. Now we will multiply three-digit decimal numbers.

Let's take 3.42 and 71.5 and do it with the help of the maths sutra vertically and crosswise.

$$
\begin{array}{r}
3.42 \\
\times\ 71.5 \\
\hline
\end{array}
$$

Step 1

Like before, we first ignore the decimal points and solve the sum. We will put the decimal point later in our final step. We multiply vertically first. So we multiply 2 x 5, which gives us 10. We put down 0 and carry 1 over to the next step.

$$
\begin{array}{r}
3.42 \\
\times\ 71.5 \\
\hline
10
\end{array}
$$

Step 2

In the second step, we multiply crosswise. So we have (5 x 4) + (1 x 2) = 22 + 1 (carry-over) = 23. We put down 3 and carry 2 over to the next step.

$$
\begin{array}{r}
3.42 \\
\times\ 71.5 \\
\hline
2310
\end{array}
$$

Step 3

We now do (5 x 3) + (7 x 2) + (1 x 4) = 33 plus the carry-

over 2, which gives us 35. We put down 5 and carry over
3. Our sum looks like this now:

$$
\begin{array}{r}
3.42 \\
\times\, 71.5 \\
\hline
352310 \\
\end{array}
$$

Step 4

We now come to the second last step.

Here we multiply crosswise. So we have $(1 \times 3) + (7 \times 4) =$
31 plus the carry-over 3, which gives us 34.

We put down 4 and carry over 3 to the next step.

$$
\begin{array}{r}
3.42 \\
\times\, 71.5 \\
\hline
34352310 \\
\end{array}
$$

Step 5

In our final step, we multiply vertically. We have 7×3 giving
us 21, plus the 3 carried over, which gives us 24. We put
down 24 and our answer becomes 244530.

$$
\begin{array}{r}
3.42 \\
\times\, 71.5 \\
\hline
2434352310 \\
\end{array}
$$

We now count and put our decimal point.

In 3.42, the decimal point is two places from the right, and
in 71.5, the decimal point is one place from the right. So we
have $2 + 1 = 3$, which means that in the final answer, the
decimal is put three places from the right. So our complete
answer becomes 244.530.

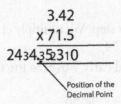

Position of the
Decimal Point

Let's try another sum 2.38 x 9.01.

$$\begin{array}{r} 2.38 \\ \times\, 9.01 \\ \hline \end{array}$$

We first ignore the decimal points and solve the sum accordingly by the vertically and crosswise pattern for three-digit numbers.

Step 1
We multiply 1 x 8 vertically. This gives us 8 as the unit's digit of our answer.

$$\begin{array}{r} 2.38 \\ \times\, 9.01 \\ \hline 8 \end{array}$$

Step 2
We multiply crosswise now (1 x 3) + (0 x 8) = 3. We put down 3 as our ten's digit.

$$\begin{array}{r} 2.38 \\ \times\, 9.01 \\ \hline 38 \end{array}$$

Step 3

Now we do the star step. We multiply $(1 \times 2) + (9 \times 8) + (0 \times 3) = 2 + 72 + 0 = 74$.

We put down 4 and carry 7 over to the next step.

$$\begin{array}{r} 2.38 \\ \times\, 9.01 \\ \hline {}_7438 \end{array}$$

Step 4

We again multiply crosswise $(0 \times 2) + (9 \times 3) = 27 + 7$ (carried over) $= 34$. We then put down 4 and carry 3 over to the next step.

$$\begin{array}{r} 2.38 \\ \times\, 9.01 \\ \hline {}_34_7438 \end{array}$$

Step 5

Finally, we multiply vertically again. $(9 \times 2) = 18 + 3$ (carried over) $= 21$.

So we get 214438.

We now have to place the decimal point. We see that in both the numbers the decimal point is two places from the right. So we add them and we put the decimal point four places from the right.

$$\begin{array}{r} 2.38 \\ \times\, 9.01 \\ \hline 21._34_7438 \end{array}$$

Position of the
Decimal Point

Our final answer is 21.4438.

I hope you have understood this concept of putting the decimal places. It's quite simple and easy. We just have to first solve the sum ignoring the decimals and then put back the decimal point after counting as was shown in the earlier sums.

Let's now move on to division of decimals.

Divisions by 10, 100, 1000, Etc.

The process of division by powers of ten is exactly the opposite of the process of multiplication by powers of ten. When we divide by a power of ten, we move the decimal point to the left.

The numbers of places by which you move the decimal point depends on the number of zeroes in the power of ten.

Let's divide 3.17 by 10. Since 10 has one zero, just move the decimal point one place to the left to make it 0.317.

Similarly, let's divide 4.52 by 100. We see that 100 has two zeroes. So we just move the decimal point two places to the left to make it 0.0452.

Below are examples of what happens to a number when it's divided by different powers of ten.

Number	÷ 10	÷ 100	÷ 1000	÷ 10000
0.73	0.073	0.0073	0.00073	0.000073
0.891	0.0891	0.00891	0.000891	0.000089
8.432	0.8432	0.08432	0.008432	0.000843
657.745	65.7745	6.57745	0.657745	0.065775
832.901	83.2901	8.32901	0.832901	0.08329

6.7234	0.67234	0.067234	0.006723	0.000672
6894.942	689.4942	68.94942	6.894942	0.689494
93.05	9.305	0.9305	0.09305	0.009305
67823.437	6782.344	678.2344	67.82344	6.782344

Dividing Decimals

Dividing a decimal by a whole number

Step 1
We first divide ignoring the decimal point.

Step 2
We then put the decimal point in the same place as the dividend.

Let's try $9.1 \div 7$.

Step 1
We first solve this ignoring the decimal places, $91 \div 7 = 13$.

Step 2
We now put the decimal point in our answer in the same place as the dividend. So our answer becomes 1.3.

Let's take another example $5.26 \div 2$.

Step 1
We first solve this ignoring the decimal places, $526 \div 2 = 263$.

Step 2

We now put the decimal point in our answer, right in the same place as it was in the dividend. So our answer becomes 2.63.

Isn't it simple and easy?

Dividing a decimal by another decimal

So what do we do when we want to divide by another decimal number?

The trick is to convert the number you are dividing by to a whole number first, by shifting the decimal point of both numbers to the right.

For example, if you have $567.29 \div 45.67$.

Step 1

We shift the decimal point of the divisor two places to the right, and it becomes 4567. Similarly, we shift the decimal point two places to the right in the dividend and it becomes 56729.

So our sum becomes $56729 \div 4567$.

Step 2

We divide by the flag method and get our answer as 12.4215.

Let's take another sum to understand it a little better.

Let's consider $7.625 \div 0.923$.

Step 1

We first shift the decimal point of the divisor three places to the right to make it a whole number 923. Similarly, we shift the decimal point of our dividend three places to the right. It becomes 7625.

So our sum becomes 7625 ÷ 923.

Step2

We divide normally and get our answer as 8.2611. Yes it's that simple.

We just have to learn to play around with the decimal point.

Negative Expectancy

Sometimes, during a maths exam, when you see a sum that scares you off, this puts you in pessimistic mode. You tell yourself over and over again that you will not be able to solve

or even attempt this sum. Such a situation is called expecting the negative or negative expectancy.

An inferiority complex and extreme lethargy are sure symptoms of negative expectancy. Extremely lethargic people always feel like lying down all the time without working and they just want to be left alone. This kind of mindset and attitude totally affects your success rate in a maths exam and also in life.

So how do we overcome this problem of negative expectancy or pessimism towards a mathematics examination?

1. Well, we can begin by dispelling the negative energy surrounding us using a smile. We can smile when we sit down to study mathematics. An improvement in our inner self will certainly improve our outer self.
2. Another way of fighting this state of mind is to be good at some maths topic and excel in it. You should be so good at this topic that you can solve it even blindfolded. This will give you a great boost of positive energy.
3. And finally, run. Running is very therapeutic whenever we are feeling negative. Just jog for 15–20 minutes! Don't stop till you're sweating profusely! After you are done with your jog, start studying straight away. This will propel you in the right direction.

8

Recurring Decimals

'Success is the sum of small efforts, repeated day in and day out.'

— Robert Collier

YES MY FRIEND, success is the sum of small efforts you take. It's just like a piggy bank—you put coins in the piggy bank every day and then it all collects to a huge sum. Similarly, success is also the sum of small efforts over time. One day, you realize you have put in so much of hard work and perseverance that success just seems to be the by-product.

In this chapter, we shall look at methods to convert fractions to decimals.

Three Types of Decimals

There are actually three types of decimals that we come across.

1. Recurring decimals

2. Non-recurring decimals
3. Non-recurring and unending decimals

Let's spend some time and learn about each of them.

1. *Recurring decimals*

Let's first take the case of recurring decimals, which basically consist of never-ending digits that repeat or recur.

A few examples:

$$\frac{1}{3} = 0.333\ldots$$

$$\frac{1}{9} = 0.1111\ldots$$

Sometimes, the digits repeat in groups like this,

$$\frac{1}{99} = .01010101\ldots$$

These decimals occur whenever the denominator of the fraction has prime numbers other than 2 or 5, such as 3, 7, 11 and 13 as factors.

2. *Non-recurring decimals*

A non-recurring decimal occurs whenever the denominator of a fraction has 2 or 5 as factors. Unlike recurring decimals, these terminate after a certain number of digits. A point to note here is that every 2, 5 or 10 in the denominator gives rise to one significant digit in the decimal.

For example, $\frac{1}{5} = 0.2$, $\frac{1}{2} = 0.5$, and $\frac{1}{10} = 0.1$ all terminate after one significant digit in the decimal.

Similarly, $\frac{1}{4} = 0.25$, $\frac{1}{25} = 0.4$, and $\frac{1}{100} = 0.01$ all terminate after two significant digits in the decimal.

3. *Non-recurring and unending decimals*

Let's now understand a non-recurring and unending decimal. This basically occurs when the numbers are irrational.

This occurs for example in irrationals such as

$\sqrt{2} = 1.41421356\ldots$ or constants like $\pi = 3.141592653\ldots$

Vedic One-Line Method

The conventional practice of converting a reciprocal or fraction into a decimal number involves the division of the numerator by the actual denominator. When the denominator is an odd prime number such as 19, 23, 29, 17 etc., the actual division process becomes too difficult; whereas, if we use the Vedic method, the whole division process becomes oral and the decimals of such fractions can be written directly.

Reciprocals of Numbers Ending in 9

We will first consider reciprocals of numbers ending in 9 such as 19, 29, 39, 79, etc.

Usually, dividing by 19, 29 or 39 is not too easy, but here, we will apply the maths sutra — By One More than the One Before. This basically means one more than the number before the 9.

There is a 1 before the 9, in the case of 19; one more than this is 2. So 2 is our Ekadhika which is 'one more'.

For 29, the Ekadhika is 3, because we drop nine and raise 2 by one more, which gives 3.

Similarly, the Ekadhika for 59 is 6.

Since only 9 is dropped, we replace the decimal of the numerator by shifting the decimal one place to the left. So it will become 0.1. Now the process will be to divide 0.1 by the Ekadhika number.

For example, with 29, the Ekadhika is 3. So our starting point will be 0.1 divided by 3, which can be orally done instead of remembering the table of 29 for $1 \div 29$.

Let's see the division method that will give us the answer from left to right.

1. Convert the fraction $\dfrac{1}{19}$ to its decimal form.

    Step 1
    We find the Ekadhika of $\dfrac{1}{19}$. After dropping the 9 of 19, we get one more than 1, which is 2.

    So this becomes $\dfrac{1}{19} \approx \dfrac{1}{20} \approx \dfrac{0.1}{2}$.

Step 2

Here the starting point is $0.1 \div 2$. As we cannot divide 1 by 2, we give a decimal point in the answer with 0 as quotient and remainder 1 to be written. This means our next dividend is 10.

$$\frac{1}{19} = 0._1 0 \ldots$$

Step 3

We divide 10 by 2 which gives us 5 and remainder zero. The sum looks like this:

$$\frac{1}{19} = 0._1 0_0 5 \ldots$$

Step 4

The quotient obtained is 05. We divide 05 by 2. This gives us our next quotient digit of 2 and remainder of 1.
We prefix 1 before 2 as shown below.

$$\frac{1}{19} = 0._1 0_0 5_1 2 \ldots$$

Step 5

We keep on dividing by 2 and writing down the quotients and remainders as it has already been shown. And this is what we get:

$$\frac{1}{19} = 0._1 0_0 5_1 2_0 6_0 3 \ldots$$

Our answer becomes 0.05263.

You will note that if this process of division is continued, the same set of digits will start repeating. We stop dividing once we get the required number of decimal places.

2. Convert the fraction $\frac{1}{29}$ to its decimal form.

Step 1
The Ekadhika of 29 is 3. So our starting point will be $0.1 \div 3$.

$$\frac{1}{29} \approx \frac{1}{30} \approx \frac{0.1}{3}$$

Step 2
As we cannot divide 1 by 3, we give a decimal point in the answer with 0 as quotient and remainder 1 to be written. This means our next dividend is 10.

$$\frac{1}{29} = 0._10\ldots$$

Step 3
We divide 10 by the Ekadhika 3. This gives us the quotient digit of 3 with remainder 1. It looks like this:

$$\frac{1}{29} = 0._10_13\ldots$$

Step 4
We continue our division process by the Ekadhika which is 3.

We now divide 13 by 3, which gives us our next quotient digit 4 and remainder 1.

Our sum now looks like this:

$$\frac{1}{29} = 0._10_13_14\ldots$$

Step 5

We continue the same process and get the following results.

$$\frac{1}{29} = 0._10_13_14_24_08_22\ldots$$

Our answer here is 0.034482. You will note that if this process of division is continued, the same set of digits will start repeating.

Let's now see what happens in the cases of $\frac{1}{39}$ and $\frac{1}{49}$.

$$\frac{1}{39} \simeq \frac{1}{40} \simeq \frac{0.1}{4} \simeq 0._10_22_25_1641$$

$$\frac{1}{49} \simeq \frac{1}{50} \simeq \frac{0.1}{5} \simeq 0.0204081632\ldots$$

Reciprocals of Numbers Ending in 3

12, 23 and 33 are some of the numbers ending with 3. To find out their reciprocals, we have to ensure that the denominator ends in a 9. To ensure that the denominator ends in a 9, we multiply the numerator and the denominator by 3.

Suppose we have to find out the value of $\frac{1}{13}$.

$$\frac{1}{13} = \frac{1}{13} \times \frac{3}{3} \simeq \frac{3}{40} \simeq \frac{0.3}{4}$$

We can now proceed with the same method we used with the numbers ending in 9.

So we have

$$\frac{0.3}{4} = 0.0_2 7_3 6_0 9_1 2_0 3....$$

Let's take another example $\frac{1}{23}$.

Step 1

We multiply the numerator and the denominator by 3 and make it

$$\frac{1}{23} = \frac{3}{69} \approx \frac{0.3}{7}$$

Step 2

Now, by the Vedic method, we will obtain the decimal form by dividing 0.3 by 7.

$$\frac{1}{23} = \frac{3}{69} = \frac{0.3}{7} \quad 0._3 0_2 4_3 3_5 4....$$

Reciprocals of Numbers Ending in 7

7, 17 and 27 are some of the numbers ending in 7. The process would be the same. We have to ensure that our denominator ends in a 9. The only way this is possible if we multiply the numerator and the denominator by 7.

Let's consider $\frac{1}{7}$.

Step 1

We multiply the numerator and denominator by 7 to ensure that the denominator ends in a 9.

We then apply the division method and retrieve the decimals.

So we have $\frac{1}{7} = \frac{1}{7} \times \frac{7}{7} = \frac{7}{49}$.

Step 2

Here our Ekadhika for 49 is 5.

$$\frac{1}{7} = \frac{1}{7} \times \frac{7}{7} = \frac{7}{49} \simeq \frac{7}{50} \simeq \frac{0.7}{5}$$

Step 3

Now we divide just like we had been doing previously. So we have

$$\frac{1}{7} = \frac{0.7}{5} = 0._2 1_1 4_4 2_2 8_3 5_0 7....$$

And this is our final answer.

Let's take another fraction, $\frac{1}{17}$.

Just as in the case of, we multiply the numerator and the denominator by 7 so that the denominator ends in a 9.

Step 1

$$\frac{1}{17} = \frac{1}{17} \times \frac{7}{7} = \frac{7}{119} = \frac{0.7}{12}$$

Step 2

$$\frac{1}{17} \approx \frac{0.7}{12} = 0._70_{10}5_98_28_42....$$

And this is our final answer.

Similarly, we can go on and do other reciprocals as well.

So, in this chapter, we saw how to calculate reciprocals of numbers ending in 9, 3 and 7. We used the Vedic One-Line method to solve our answers and we saw how easy it was to calculate the recurring decimals from the given reciprocals.

9

Percentages

WHAT IS A PERCENTAGE? A percentage is a way of expressing a number, especially a ratio, as a fraction of 100. The word percentage is derived from a Latin word *per centum*, which means 'by the hundred'. Percentages are denoted by %.

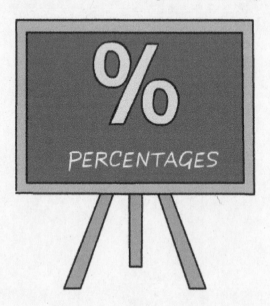

So if I say 27%, this means the fraction is $\frac{27}{100}$ or 0.27 in decimal form, which we get by dividing 27 by 100.

Supposing the percentage is 60%, then this means that the fraction is $\frac{60}{100}$ or 0.6 in decimal form, which we get by dividing 60 by 100.

Percentages are generally used to express how big or small one quantity is when comparing it with another quantity.

Now that we have some understanding of what percentages are, let's move on and see how we can convert percentages into fractions.

Converting Percentages into Fractions

Convert 75% into a fraction.

We first convert 75% into a fraction like this, $75\% = \frac{75}{100} = \frac{3}{4}$.

We have also reduced the fraction $\frac{75}{100}$ to its simplest form, which is $\frac{3}{4}$.

Let's take another sum. Convert 40% into a fraction.

So we have $40\% = \frac{40}{100} = \frac{2}{5}$. Here we have reduced $\frac{40}{100}$ to its simplest form, which is $\frac{2}{5}$.

Let's take another one: convert 65% into a fraction.

We simply write $65\% = \frac{65}{100} = \frac{13}{20}$. Here again, we have reduced the fraction $\frac{65}{100}$ to its simplest form, $\frac{13}{20}$.

Now, let's move on to see how we can convert fractions into percentages.

Converting Fractions into Percentages

Now we do just the opposite—we convert a fraction into a percentage by multiplying by 100.

Say we have to convert $\frac{3}{4}$ into a percentage.

So this is what we do:

$$\frac{3}{4} \times 100 = 3 \times 25 = 75\%$$

So here $\frac{3}{4}$ is equal to 75%.

Now say we have to convert $\frac{2}{5}$ into a percentage.

So all we have to do is multiply the fraction by 100.

$$\frac{2}{5} \times 100 = 2 \times 20 = 40\%$$

Let's take another example. Say we have to convert $\frac{17}{20}$ into a percentage.

Just like we have been doing, we multiply this fraction by 100 like this:

So the fraction $\frac{17}{20}$ is equal to 85%.

Now let's see how to convert a percentage into a decimal.

Converting Percentages into Decimals

Converting percentages into decimals is the easiest.

Say we have to convert 45.6% into a decimal.

So what we do is shift the decimal point two places to the left. So our answer is:

$$\frac{45.6}{100} = 0.456$$

Let's take another example: convert 8.09% into a decimal. This is the same as dividing 8.09 by 100. So we shift our decimal point two places to the left like this:

$$\frac{8.09}{100} = 0.0809$$

So our answer is 0.0809.

Let's take another example: convert 0.674% into a decimal. Again we do the same thing—we divide 0.674 by 100 and shift the decimal two places to the left, like this:

$$\frac{0.674}{100} = 0.00674$$

That's about it. But now let's see how to find the percentage of a given quantity.

Finding the Percentage of a Given Quantity

Let's take an example. Find 45% of 81.

In order to find the percentage of a given quantity, multiply the percentage as a fraction by the given quantity.

So we get $\frac{45}{100} \times 81$.

Now to multiply 45 x 81 we use the popular maths sutra, vertically and crosswise. Let's do a quick recap of the formula.

2-Digits Vertically & Crosswise Pattern

Step 1

The dots represent numbers. So we first multiply vertically as shown by the arrows.

Step 2

We then multiply crosswise as shown by the arrows.

Step 3

Finally, we multiply vertically again as shown by the arrows. I am sure by now you must remember it. So back to the sum, we have

$$\frac{45}{100} \times 81 = \frac{3645}{100} = 36.45$$

Therefore, 45% of 81 is 36.45.

Let's go for another example: find 73% of 98.

Simply multiply 73 and 98 using the vertically and crosswise formula. Applying this formula, we get 7154, and then we simply divide by 100. We get 71.54, which is our answer.

$$\frac{73}{100} \times 98 = \frac{7154}{100} = 71.54$$

Say we have to calculate 23% of 67.

We can just multiply 23 and 67 by the vertically and crosswise method. Doing that gives us 1541. We now just have to divide by 100. We get 15.41 and this is our answer.

The sum solved looks like this:

$$\frac{23}{100} \times 67 = \frac{1541}{100} = 15.41$$

So far we have been working with two digits, but there can be a sum like this: 14.2% of 682. Here we multiply 142 by 682 and then put the decimal points.

We multiply using the vertically and crosswise method, three-digit by three-digit pattern. Let's do a quick recap and check it out.

3-Digit Vertically & Crosswise Pattern

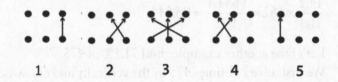

This is the three-digit by three-digit pattern. You should note here that the dots represent numbers.

Step 1
We multiply vertically.

Step 2
We then multiply crosswise and add.

Step 3
We then do the 'star' multiplication.

Step 4

Again, we multiply crosswise.

Step 5

Finally, we multiply vertically.

We are not going into the detail as we have already covered this before, so now let's get back to our sum.

Find 14.2% of 682.

We multiply 142 and 682 by the vertically and crosswise method. We get 96844. We adjust for the decimal point and it becomes 9684.4. We now divide by 100 and get 96.844.

Our sum looks like this:

$$\frac{14.2}{100} \times 682 = \frac{9684.4}{100} = 96.844$$

Let's take another example: find 71.1% of 475.

We first solve 711 times 475 by the vertically and crosswise method. This gives us 337725. We then adjust for the decimal point, which makes it 33772.5.

Now we simply divide by 100. This gives us 337.725. Our sum looks like this:

$$\frac{71.1}{100} \times 475 = \frac{33772.5}{100} = 337.725$$

Next, find 87.2% of 584.

As we have done before, we multiply 872 by 584 using the vertically and crosswise method. We get 509248. We adjust for the decimal point and make it 50924.8

Now all we have to do is to divide by 100. This gives us 509.248. Our sum looks like this:

$$\frac{87.2}{100} \times 584 = 509.248$$

Expressing One Quantity as a Percentage of Another

Find 75 as a percentage of 250.

All we have to do is divide 75 by 250 and then multiply by 100.

$$\frac{75}{250} \times 100 = 30\%$$

Find 82 as a percentage of 450.

So just like before, we divide 82 by 450 and then multiply by 100.

$$\frac{82}{450} \times 100 = 18.22\%$$

Find 35 as a percentage of 890.

We divide 35 by 890 and then multiply by 100.

$$\frac{35}{890} \times 100 = 3.932\%$$

Approximating Percentages

Sometimes, in competitive examinations, we may need to get an idea and estimate what percentage one number is of another.

For example, estimate 42 as a percentage of 800.

Here we need to be able to calculate 10% of 800 first, which is 80. Half of that or 5% of 800 is 40. Since 42 is a little more than 40, we can conclude that it is a little more than 5%. So we can say 42 is about 5% of 800.

Another example: estimate 45 as a percentage of 650.

Here we first calculate 10% of 650, which is 65. So since 45 is less than 65, we go ahead and find out what is 5% of 650. The answer is 32.5.

So here 45 is obviously more than 5%, so we go ahead and find 1% of 650, which is 6.50.

5% ——— 32.5
1% ——— 6.5
1% ——— 6.5

We find that 7% is about 45.5. So, 45 is a little less than 7% of 650.

Let's do one more sum to understand this better. Estimate 73 as a percentage of 3568.

We find 1% of 3568. This gives us 35.68. Since 73 is a little more than double of 35, we find 2% of 3568, which is 71.36. So 73 is about 2%. Estimating answers leads to intelligent guessing in competitive examinations and let us choose options in a better and faster way!

Percentage Increase or Decrease

Here, keep in mind that a percentage increase or decrease is always in relation to an original size or value.

So, supposing we have to increase the value by 18%, then instead of finding 18% and adding it to the original number,

we'll simply multiply the original number by 1.18 and get
our answer.

Say we have to increase 673 by 23%. What we need to do
is multiply 673 by 1.23 to get our answer.

Here, we'll multiply using the vertically and crosswise
method and I know for sure that all of you have become
experts at it!

Step 1
We have to multiply 673 x 1.23. Ignoring the decimals in the
beginning, we will multiply 3 x 3 to get the unit's place 9.

Step 2
We will now multiply crosswise. So we have (3 x 7) + (3
x 2) = 21 + 6 = 27. Here, we will put 7 down in the tens
place and carry over 2 to the next step.

Step 3
We will now do the star multiplication step. So we multiply
crosswise (3 x 6) + (7 x 2) + (3 x 1) = 35 + 2 (carried over)
= 37. We put down 7 and carry over 3 to the next step.

Step 4
We will now do crosswise multiplication. So here we will
multiply (1 x 7) + (2 x 6) = 7 + 12 + 3 (carried over) = 22.
We put down 2 and carry over 2 to the next step.

Step 5
We will now multiply vertically 1 x 6 = 6 + 2 (carried over)
= 8.

So our answer is 82779. We will now put the decimal point two places from the right. So we have 827.79 as our final answer.

For our next example, let's increase 450 by 34%.

In this example, instead of working out the percentage in detail and then adding it, we'll simply multiply 450 by 1.34 using the vertically and crosswise pattern.

Step 1
We multiply 4 and 0 vertically. This gives us 0 as the units digit of our answer.

Step 2
We now multiply crosswise $(4 \times 5) + (3 \times 0) = 20$. Again, we put down 0 and carry over 2 to the next step.

Step 3
We now multiply in the star formation. So we have $(4 \times 4) + (1 \times 0) + (3 \times 5) = 31 + 2$ (carried over) $= 33$. We put down 3 and carry over 3 to the next step.

Step 4
We then multiply $(3 \times 4) + (1 \times 5) = 12 + 5 = 17 + 3$ (carried over) $= 20$. So put down 0 and carry over 2 to the next step.

Step 5
In our final step, we multiply vertically: $(1 \times 4) = 4 + 2$ (carried over) $= 6$. So it becomes 60300.

We put the decimal point two places to the left from the end. Our answer is 603.00.

Let's now look at a decrease in percentages.

Say we have to decrease 500 by 25%. So we'll multiply 500 by 0.75, that is, 1 - 0.25. We multiply by 0.75, because we have to find out 75% of 500 or 25% less than 500. So we solve 0.75 x 500 using the vertically and crosswise method.

Step 1
We multiply vertically 5 x 0, which gives us 0 as the units digit.

Step 2
Next, we multiply crosswise (5 x 0) + (7 x 0) = 0. This zero is the ten's place digit.

Step 3
Now we do the star step (5 x 5) + (0 x 0) + (7 x 0) = 25. We put down 5 and carry over 2 to the next step.

Step 4
We then do the crosswise step. So we have (7 x 5) + (0 x 0) = 35 + 2 (carried over) = 37. We put down 7 in the thousand's place and carry over 3 to the next step.

Step 5
We now multiply vertically (0 x 5) = 0 + 3 (carried over) = 3. This is the final step and our answer is 37500. We put a decimal point two digits from the left. So we have our answer as 375.00.

For our next example, let's decrease 878 by 62%.

Here we'll multiply 878 by 0.38, that is 1 - 0.62. We multiply

by 0.38 because we have to find out 38% of 878, or 62% less
than 878.

Step 1
We multiply vertically 8 x 8, which gives us 64. We put
down 4 and carry over 6 to the next step.

Step 2
We now multiply crosswise $(8 \times 7) + (3 \times 8) = 56 + 24 = 80$
+ 6 (carried over) = 86. So we put down 6 and carry over
8 to the next step.

Step 3
We now do the star step $(8 \times 8) + (0 \times 8) + (3 \times 7) = 64 +$
$0 + 21 = 85 + 8$ (carried over) = 93. We put down 3 and
carry 9 over to the next step.

Step 4
Now we multiply vertically! So we have $(3 \times 8) + (0 \times 7) =$
$24 + 0 = 24 + 9$ (carried over) = 33. We put down 3 and
carry over 3 to the next step.

Step 5
Finally, we multiply $(0 \times 8) = 0 + 3$ (carried over) = 3. So
it becomes 33364. We put the decimal place two places to
the left from the end and get 333.64 as our answer.

So we have 878 x 0.38 = 333.64, which is 62% less than 878.

Now for the final example of this chapter: decrease 345 by 18%.
Here we'll simply multiply 345 by 0.82, or 345 x (1 - 0.18).

Step 1
We multiply 2 x 5 = 10. We put down 0 and carry over 1 to the next step.

Step 2
We now multiply crosswise. So it becomes (2 x 4) + (8 x 5) = 8 + 40 = 48 + 1 (carried over) = 49. We put down 9 and carry 4 over to the next step.

Step 3
We now do the star step. So we have (2 x 3) + (0 x 5) + (8 x 4) = 6 + 0 + 32 = 38 + 4 (carried over) = 42. We put down 2 and carry over 4 to the next step.

Step 4
Now we multiply (8 x 3) + (0 x 4) = 24 + 4 (carried over) = 28. We keep the eight and carry over 2.

Step 5
We multiply 0 x 3 = 0 + 2 (carried over) = 2.

So our answer is 28290. We put the decimal two places to the left, which gives us 282.90.

This makes percentages very easy! All we have to do is to apply the maths sutra vertically and crosswise and then multiply! That's all there is to percentages—really!

Practising Gratitude to Boost Maths Scores

One of the best ways to achieve that perfect maths score is to perfect the art of being grateful. You must be wondering how this happens. What is the link between being thankful and your maths scores? They seem to be totally unrelated, right? Wrong!

It's easy to be thankful and show gratitude for nice things in your life, but being grateful for unpleasant, uncomfortable and frightening stuff such as maths can be challenging to begin with. No matter what the situation is, just stay focused and don't give up! If you have a tough time doing your maths homework or have a hard time understanding the concept of algebra or you failed in your maths paper no matter what, just stay positive and never give up.

What can be worse than failing in your maths paper? The next time you feel this way, just think about the millions of underprivileged children around the world who don't even get to go to school. At least this will help you be grateful towards your teachers and parents who work so hard to give you the best. So the next time you get bored or feel lazy, just think about these things and motivate yourselves.

For beginners, here's what you can do—take a new diary and start penning down things in maths that you are grateful for. You can pen down other things as well. You can write things like:

- I am grateful to my maths teacher for teaching me the concept of algebra.
- I am grateful to my maths teacher for pointing out my mistakes and showing me ways to improve my maths score.
- I am so happy and grateful to my friends for helping me find the correct answer to a difficult problem just before an exam.

Rather than just writing it, feel thankful inside. Feel it with all your heart and start reflecting or meditating about what you are thankful about. Then write it—it will give you a new perspective.

So remember, expressing thankfulness towards maths might be a little challenging. But if you are persistent, then your positivity will make an impact on your maths scores. Trust me on this!

10

Divisibility

IN THIS CHAPTER we will be learning about divisibility. We will
learn how to ascertain whether a particular number is divisible
by another number or not. This can be done through a method
called osculation. But before we launch into osculation, let us
see some known divisibility rules of some numbers.

Divisibility

How do we know if a number is divisible by 2 or not?
This is very simple. For a number to be divisible by 2, the last
digit of the given number must be even—2, 4, 6, 8 or 0.

How do we know if a number is divisible by 3 or not?
I am sure all of you know this already. To see if a number is
divisible by 3, we add all the digits in the number, and if the
sum of the digits is divisible by 3, then we can say that the
number is divisible by 3. For example, if we have the number
345, we can say that the number is divisible by 3, because 3 +
4 + 5 = 12, and 12 is a multiple of 3.

If the last two digits of a number are divisible by 4, then we can say that the number is divisible by 4. For example, if we have 8732, the last two digits are 32, which is divisible by 4; so we can say that 8732 is divisible by 4.

For a number to be divisible by 5, its last digit has to be 5 or 0. This is very simple.

Know what 6 is the first composite number, so we check for divisibility by both 2 and by 3.

For 7, we will soon see how to find the divisibility rule.

If the last three digits are divisible by 8, then we can say that the number is divisible by 8. For example, if the number is 337312, the last three digits 312 are divisible by 8. Now we can say that the number is divisible by 8.

A number is divisible by 9 if all the digits in the number add up to 9 or a multiple of 9.

Lastly, we can say that a number is divisible by 10 if the number ends in a 0.

Let us now see the divisibility rule for larger numbers—especially prime numbers.

The Osculator

The osculator is the number **one more than the one before** when a number ends in a 9 or a series of 9s.

1. For example, if we have 29, the osculator is 3, because the one before the 9 is 2 and one more than 2 is 3.

2. If we have 59, the osculator is 6, because the one before the 9 is 5 and one more than 5 is 6.

3. Likewise, the osculator for 79 is 8.

4. For 13 the osculator is 4, because to obtain a 9 at the end

we must multiply 13 by 3, which gives 39, for which the osculator is 4.

5. Similarly for 7, the osculator is 5, because to obtain a 9 at the end we must multiply 7 by 7, which gives us 49, for which the osculator is 5.

6. For 17, the osculator is 12, because to obtain a 9 at the end we must multiply 17 by 7, which gives us 119, for which the osculator is 12.

7. For 23, the osculator is 7, because to obtain a 9 at the end we must multiply 23 by 3, which gives us 69, for which the osculator is 7.

The Osculation Method

This method will help us find the divisibility rule for any number.

1. Say we have to find out if 112 is divisible by 7.

   We just learnt that the osculator for 7 is 5.

   Now we osculate 112 with 5.

   To osculate a number, we multiply its last figure by the osculator of the divisor (it is 7 here) and add the result to its previous figure.

   So, say we have 112.

   $11 + (2 \times 5)$

   $= 11 + 10$

   $= 21$

   Now we osculate again. We multiply 1 by 5 and add to 2.

   $= 2 + (1 \times 5)$

   $= 7$

Since our osculation result is the divisor itself (7), we can clearly say that 112 is divisible by 7.

2. Let's take another example. Say we want to know if 49 is divisible by 7.
   Again we find the osculator of 7, which is 5.
   We now osculate 49 with 5.

   This becomes
   49
   $= 4 + (9 \times 5)$
   $= 49$
   $= 4 + (9 \times 5)$
   $= 49$

   We see that there is a repetition of 49 as we osculate. And, whenever we osculate, we will see that only multiples of 7 are produced. And osculating any number which is not a multiple of 7 will never produce a multiple of 7. This means 49 is divisible by 7.

3. Let's take another sum. Say we have to find out if 2844 is divisible by 79.
   We first find the osculator of 79, which is 8.

   We now osculate 2844 with 8 like this:
   2844
   $284 + (4 \times 8)$
   $= 284 + 32$
   $= 316$
   $= 31 + (6 \times 8)$

$= 31 + 48$
$= 79$

Since our osculation result is the divisor itself (79), we can clearly say that 2844 is divisible by 79.

Now how easy was that! You just have to find out the osculator and then simply osculate. That's all there is to this. So now we can find out the divisibility rule for any number.

Let's see another sum. Is 1035 divisible by 23?

First, let's find the osculator. We multiply 23 by 3, which makes it 69. Now we apply the maths sutra, one more than the one before — one more than 6 is 7. Our osculator is 7.

We now osculate 1035 with 7.
1035
$= 103 + (5 \times 7)$
$= 138$
$= 13 + (8 \times 7)$
$= 13 + 56$
$= 69$
$= 6 + (9 \times 7)$
$= 69$

We see that 69 is repeating. 69 is the multiple of 23. So we can safely say that 1035 is divisible by 23.

4. Is 6308 divisible by 38?

We first see that 38 is a composite number made up of 2 and 19.

2 x 19 = 38, so here we will test for divisibility by 2 and 19. Since 6308 is an even number, we know that it is surely divisible by 2. Now we need to test for divisibility by 19. We apply the sutra, one more than the one before, and get one more than 1, which is 2. Our osculator is 2. Now we osculate 6308 with 2.

6308
$= 630 + (8 \times 2)$
$= 646$
$= 64 + (6 \times 2)$
$= 76$
$= 7 + (6 \times 2)$
$= 19$

We got 19 as the result of the osculation process, which means 6308 is divisible by 38.

5. Let's do another sum. Check for the divisibility of 334455 by 39.
We find the osculator of 39, which is 4.
Let's osculate 334455 with 4.

Our process looks like this:
334455
$= 33445 + (5 \times 4)$
$= 33465$
$= 3346 + (5 \times 4)$
$= 3366$
$= 336 + (6 \times 4)$
$= 360$

$$= 36 + (0 \times 4)$$
$$= 36$$

Since we got the final result as 36, which is below our divisor of 39, we can state that the number 334455 is not divisible by 39!

6. Is 3588 divisible by 69 or not?

I hope these sums have become easier now. The osculator of 69 is 7, which I am sure by now you know by heart. So let's quickly move on to the next step and osculate 3588 by 7.

3588
$$= 358 + (8 \times 7)$$
$$= 414$$
$$= 41 + (4 \times 7)$$
$$= 69$$

Since we got 69 after the osculation process and it is also our divisor, we can safely say that 3588 is divisible by 69.

The Negative Osculator

For a number to have a negative osculator, the divisor should end in a 1.

Say we have a divisor 51, the negative osculator would be 5—we just drop the 1.

The negative osculator for the divisor 81 would be 8, since we just drop the 1.

Say the divisor is 7. We multiply 7 by 3 to make it end in a

1, so we have 21. We now just drop the 1 and get our negative osculator 2.

Say the divisor is 9. We multiply this by 9 to ensure that it ends in a 1. So we have 81. We simply drop the 1 and get our negative osculator 8.

The Osculation Method

Let's take a sum: Is 6603 divisible by 31?
We find that the negative osculator for 31 is 3 — we simply drop the 1.

Now we osculate 6603 with 3.

To osculate a number, we multiply its last figure by the negative osculator and then subtract the result from its previous figure.

So we have:
660 - (3 x 3) = 651
65 - (1 x 3) = 62
6 - (2 x 3) = 6 - 6 = 0

For a number to be divisible, the result of the osculation should be the divisor, zero or a repetition of a previous result. This 0 indicates that the number 6603 is divisible by 31.

Let's take another sum: Is 11234 divisible by 41?

The negative osculator for 41 is 4, which we get by simply dropping the 1.

We now osculate 11234 by 4.

So we have
1123 - (4 x 4) = 1107
110 - (7 x 4) = 82

$8 - (2 \times 4) = 0$

This 0 indicates that the number 11234 is divisible by 41 completely.

Let us take another example. Find out if 2275 is divisible by 7.

We first find the negative osculator. We do this by multiplying 7 by 3, which gives us 21. We then drop 1 and get the negative osculator 2.

Now we start osculating. So we have:

$227 - (5 \times 2) = 217$

$21 - (7 \times 2) = 7$

Since we got 7 (which is also the divisor) as the result of the osculation process, we can safely say that 2275 is divisible by 7.

Check for divisibility of 464411 by 71.

We first find the negative osculator, which is 7 here.

We now start osculating:

$46441 - (1 \times 7) = 46434$

$4643 - (4 \times 7) = 4615$

$461 - (5 \times 7) = 426$

$42 - (6 \times 7) = 0$

This zero indicates that the number 464411 is divisible by 71 completely.

Points to Be Noted

1. In all cases, the sum of the positive and the negative osculators equal the divisor.
2. For divisors ending in 1 and 7, the negative osculators are smaller than the positive osculators. Since the negative osculators are smaller, we use them.

3. And for divisors ending in 3 and 9, the positive osculators are smaller, so we use them.

Harnessing Your Brain Power

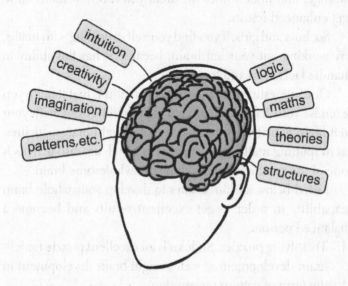

The choices you make and the actions you take determine the results you get in life, and your choices and actions stem from how you think and how you use your brain power. Unless you change how you think, you will continue to get pretty much the same results as you always have, because you probably choose the same actions and behaviours. But you can develop different ways to think and get different results by engaging different parts of your brain.

There are two parts of your brain, the left and the right. They coordinate among themselves to produce the best results. The left brain focuses on logic, maths, theories and structures, while the right brain deals with intuition, creativity, imagination, patterns, etc.

Now compare the left and the right sides of your brain by their properties and figure out which side tends to be stronger and understands the areas you can now focus on to get enhanced results.

So, boys and girls, if you find yourself struggling with maths, try working out your left brain, because it has the ability to handle facts, data and logic.

One interesting thing about Vedic maths is that it helps you exercise your left as well as your right brain. Along with your left-brain activity, you tend to do lot of right-brain activities, as in spotting mathematical patterns which are visual. This is one of the good ways to develop your wholesome brain.

Listed below are three ways to develop your whole brain capability in order to get excellent results and become a balanced person.

1. Try solving puzzles. Sudoku is an excellent puzzle for left-brain development as well as right-brain development in the form of pattern recognition.

2. Build in rest periods when you study. Study in full brain efficiency for ninety minutes or so. Then take a small break and then get back to your studies again.

3. When you set your targets in maths, make sure you do lot of self-talk, and visualize yourself doing it (right-brain activity). For example, say to yourself, 'I am the best . . . I have scored 100 per cent in mathematics,' which will keep you motivated.

11

Squares

ONCE YOU LEARN the maths sutras you will see that there are many methods to square numbers. What I am going to share with you now is the general method to square numbers. It is called the Squaring by the Duplex method and it is an application of yet another vertically and crosswise method, which you will eventually learn.

This is an important topic from the perspective of competitive examinations. In this chapter, you will learn to square not only from right to left but also mentally in one line. Let's get started.

The Duplex

As the name suggests, duplex simply means dual or something relating to two. We will be using this concept effectively to find out the squares of different numbers. The role of the duplex is to find the squares of numbers.

There are three types of duplexes:

Duplex of Individual Digits

Duplex of individual digits (single-digit figures) simply refers to its square or a^2.

So the duplex of 7 will be its square 49.

The duplex of 8 will be its square, which is 64.

And similarly, the duplex of 5 will be 25.

Do keep in mind that this is valid only for individual digits.

We will now look at the second type of duplex.

Duplex of Numbers with Even Digits

Here we are not talking about even numbers, but about numbers with even digits, meaning those numbers where the number of digits is even.

For example, let's take the number 73. This is an even-digited number because the number of digits in the figure is two, 7 and 3.

64 is an even-digited number as well, simply because it has two digits—6 and 4.

Now to find the duplex of an even-digited number we need to apply the formula $2ab$, where a and b refer to the first and second digits respectively.

Suppose that we have to find the duplex of 81, it will be $2 \times 8 \times 1 = 16$.

So the duplex of 81 is 16.

Let's find the duplex of 73, which will be $2 \times 7 \times 3 = 42$.

The duplex of 1234 will be $(2 \times 1 \times 4) + (2 \times 2 \times 3) = 8 + 12 = 20$.

Let's find the duplex of $8231 = (2 \times 8 \times 1) + (2 \times 2 \times 3) = 16 + 12 = 28$.

Let's understand this with another example. The duplex of 7351 will be $(2 \times 7 \times 1) + (2 \times 3 \times 5) = 14 + 30 = 44$.

I hope this is clear.

Now let us learn how to find the duplex of odd-digited numbers.

Duplex of Odd-Digited Numbers

The duplex of odd-digited numbers is a combination of the duplex of individual digits and the duplex of even-digited numbers—a hybrid of the two duplexes we just learnt.

For example, let us take the number 178. This is an example of an odd-digited number, because the number of digits in the figure is three, which is odd.

Now to find the duplex of an odd-digited number we'll apply the formula $m^2 + 2ab$.

Here m is for middle, and a and b are the first and the last digits respectively.

Now let us find out the duplex for 372.
$372 = 7^2 + (2 \times 3 \times 2) = 49 + 12 = 61$

Next, the duplex for 286 will be
$286 = 8^2 + (2 \times 2 \times 6) = 64 + 24 = 88$

And the duplex for 789 will be
$789 = 8^2 + (2 \times 7 \times 9) = 64 + 126 = 190$

Now that we have understood the concept of duplex, finding squares will be real fun and easy! Let us apply this and find out the square of any number.

Finding Squares Using the Duplex Method

Two-digit squares

2-Digit Squaring

$(ab)^2$ = Duplex of (a|ab|b)

Let's start with our first example: 57^2.
Here $a = 5$ and $b = 7$.

Step 1

We first find the duplex of 5, which is 25.
Then we'll find the duplex of 57 using the formula $2ab$,
which is $2 \times 5 \times 7 = 70$.
And lastly, we'll find the duplex of 7, which is 7^2 or 49.
We write the duplexes like this: 25|70|49

Step 2

In this step, we'll start adding from right to left.

25|70|49

3249

- We put down 9 from 49 and carry over 4 to the next step.
- We then add $70 + 4$, which gives us 74. Later, we'll place 4 down and carry over 7 to the next step.
- Finally, we add $25 + 7$ (carried over), so this gives us 32.
- Our answer is 3249.

Let's take another illustration. Say we have 74^2.

Step 1

We first find out the duplexes of all the digits.

- So the duplex of 7 will be 7^2, which is 49.
- Next, we'll find out the duplex of 74 and that is 2 x 7 x 4 = 56.
- Finally, we'll find the duplex of 4, which is 16.

Now all we have to do is place these duplexes down like this: 49|56|16

Step 2

Now in this step, we'll start adding to get to the final answer. So let's start from right to left.

- First, we put down 6 from 16 as the first answer digit and then carry over 1 to the next step.
- So 56 + 1 = 57. We now put down 7 as the next answer digit and carry over 5 to the next step.
- So finally, 49 + 5 = 54.

With this, our complete answer is 5476.

I'm sure that, slowly and gradually, with each example, you are gaining more and more confidence. Now let's move on to three-digit squares.

Three-digit squares

3-Digit Squaring

$(abc)^2$ = Duplex of (a|ab|abc|bc|c)

Once you start solving more practice exercises, it will be really easy. Take my word for it!

For our first three-digit example, let's take 746^2.

We now calculate the duplexes from right to left. We need to remember here that each column will give us only one digit, just like before.

Step 1

We'll first find all the duplexes and write them down. We will find the duplex of a, ab, abc, bc and c.

- The duplex of 7 is 7^2, which is 49.
- The duplex of $74 = 2 \times 7 \times 4 = 56$.
- The duplex of $746 = 4^2 + (2 \times 7 \times 6) = 16 + 84 = 100$.
- The duplex of $46 = 2 \times 4 \times 6 = 48$.
- The duplex of $6 = 6^2 = 36$.

The duplexes for 746^2 looks like this:
49|56|100|48|36

Step 2

We need to remember here that each column will give us only one digit.

- First we put down 6 from 36 as the unit's digit and carry over 3 to the next step.
- We then add 3 to 48, so it becomes 51. Then we put down 1 and carry over 5 to the next step.
- $100 + 5 = 105$; we now put down 5 and carry over 10.
- $56 + 10 = 66$. We repeat the same thing; we put down 6 and carry over 6 to the last step.
- So $49 + 6 = 55$ and finally, we take 55 down.

49|56|100|48|36

556516

Our answer is 556516.

I hope this is clear. Finding the duplexes is one thing and adding is another. We have to be adept at both and then finding the squares will be child's play.

Let's quickly solve another example: 357^2

Step 1

As usual we put down all the duplexes.

- So the duplex of 3 is 3^2, which is 9.
- The duplex of 35 is $2 \times 3 \times 5 = 30$.
- The duplex of 357 = $5^2 + (2 \times 3 \times 7) = 25 + 42 = 67$.
- The duplex of 57 = $2 \times 5 \times 7 = 70$.
- The duplex of 7 is 7^2 or 49.

Now we place all the duplexes like this:
9|30|67|70|49

Step 2

Now all we have to do is add up these duplexes.

- We'll put down 9 of 49 in the unit's place and then carry over 4 to the next column.
- So 70 + 4 (carried over) = 74; we will put the 4 down in the ten's place and carry over 7 to the next column.
- Now 67 + 7 = 74 again. So we'll put 4 down in the hundred's place and carry over 7 to the next column.

- $30 + 7 = 37$. We'll place 7 in the thousand's place and carry over 3 to the next column.
- And finally, $9 + 3 = 12$, which brings us to our final answer 127449.

So, from now on, there is no need for those complicated multiplications to find the square of a number. With this method, finding squares of numbers will be as easy as pie. We only need to take care that we have the correct duplex and that our additions are right!

Four-digit squares

4-Digit Squaring

$$(abcd)^2 = \text{Duplex of } (a|ab|abc|abcd|bcd|cd|d)$$

Let's take an example: 2894^2

Step 1
We'll find the duplexes starting from left to right:
- The duplex of 2 is 2^2, which is 4.
- The duplex of $28 = 2 \times 2 \times 8 = 32$.
- The duplex of $289 = 8^2 + (2 \times 2 \times 9) = 64 + 36 = 100$.
- The duplex of $2894 = (2 \times 2 \times 4) + (2 \times 8 \times 9) = 16 + 144 = 160$.
- The duplex of $894 = 9^2 + (2 \times 8 \times 4) = 81 + 64 = 145$.
- The duplex of $94 = 2 \times 9 \times 4 = 72$.
- The duplex of 4 = 4^2 or 16.

Now we'll arrange the duplexes and then add them up in the next step.
4|32|100|160|145|72|16

Step 2

2894^2

4|32|100|160|145|72|16

8375236

Now pay close attention. It's the same method as the one used in the earlier examples. We'll start from right to left. Do remember that each column will give us only one digit. 4|32|100|160|145|72|16

- First, we'll put down 6 from 16 and carry one to the ten's place. So our unit's digit is 6.
- 72 + 1 (carried over) = 73. Now we'll place 3 down and carry over 7 to the next column. So our ten's digit is 3.
- 145 + 7 (carried over) = 152. Here we'll put down 2 and carry over 15 to the next column.
- 160 + 15 (carried over) = 175. We'll now put down 5 and carry over 17 to the next column.
- 100 + 17 (carried over) = 117. Now we'll place 7 down and carry over 11 to the next column.
- 32 + 11 (carried over) = 43. So we now need to take down 3 and carry over 4.
- Finally, 4 + 4 (carried over) = 8. So our final answer is 8375236.

Let's take another example and square 1234^2.

Step 1

We'll first find out all the duplexes for the number 1234.
- So the duplex of 1 is $1^2 = 1$.
- Next, we'll find the duplex of 12, which is $2 \times 1 \times 2 = 4$.

- The duplex of $123 = 2^2 + (2 \times 1 \times 3) = 4 + 6 = 10$.
- Now the duplex of 1234 is $(2 \times 1 \times 4) + (2 \times 2 \times 3) = 8 + 12 = 20$.
- The duplex of $234 = 3^2 + (2 \times 2 \times 4) = 9 + 16 = 25$.
- Next, we'll find the duplex of 34, which is $2 \times 3 \times 4 = 24$.
- Finally, the duplex of 4 is 4^2 or 16.

So we have the duplexes as follows: 1|4|10|20|25|24|16

Step 2

This step is always about addition.

We'll start from the right to the left. Do remember that each column will give us a single digit.

1234^2

1|4|10|20|25|24|16

1522756

- So we'll place the 6 of 16 in the unit's place and we'll carry 1 over to the next column.
- $24 + 1$ (carried over) = 25. We then place 5 down in the ten's place and carry over 2.
- $25 + 2$ (carried over) = 27. We now place 7 and carry over 2 to the next column.
- $20 + 2$ (carried over) = 22. So we'll now place 2 down and carry over 2 to the next column.
- $10 + 2$ (carried over) = 12, so the cycle goes on. We'll now put down 2 and carry over 1 to the next step.
- So $4 + 1$ (carried over) = 5. We now place this 5 down and, voila, there is nothing to carryover!
- Now bring down 1.

And there you have it, our answer is 1522756.

Finding squares was never so easy. I remember sitting in the exam hall scratching my head, doing multiplication after multiplication to find out a single square! But I am sure that, with the duplex method, your life will be a lot easier when it comes to calculating squares.

Making Squares Easier

The duplex method that we learnt here can be turned into an even easier method. Let me show you how.

Let's take a two-digit number, say 73^2.

Step 1
Find the duplex of 3 mentally—it is 9. Now place 9 in the unit's place.

Step 2
Try to find the duplex of 73 mentally—it is 42. So now we place 2 down and carry 4 over.

Step 3
Next, we need to find the duplex of 7 and that is 49. 49 + 4 = 53.

So our answer is 5329. Wasn't that really quick? That's about it!

So now, let's solve our final example: 86^2.

Step 1
As usual, we'll first find the duplex of 6 and that's 36. Now we'll place 6 down and carry 3 forward.

Step 2
And the duplex of 8 and 6 is 96, and adding 3 gives us 99. Now we'll place 9 down and carry over 9 to the next column.

Step 3
Finally, duplex of 8 is 64, and adding the carry-over 9 gives us 73. Our complete answer is 7396.

Remember this can be extended for three digits, four digits and above.

12

Cubes

FIRST, WE NEED to understand the basic formula $(a + b)^3 =$
$a^3 + 3a^2b + 3ab^2 + b^3$

This formula is traditionally used to find the cube of any
number. Here we will expand this formula further in two

lines, which would make the calculation simpler and easier to understand.

In the first line, we write $a^3 + a^2b + ab^2 + b^3$ and in the second line, we write $2a^2b + 2ab^2$.

So if we add both the lines together we get the expression $(a + b)^3$, which is $a^3 + 3a^2b + 3ab^2 + b^3$.

This is how our expansion will look:

$$(a+b)^3 = \begin{array}{r} a^3 + a^2b + ab^2 + b^3 \\ +2a^2b + 2ab^2 \\ \hline a^3 + 3a^2b + 3ab^2 + b^3 \end{array}$$

If you look at it closely, you'll be able to spot a ratio between a^3 and a^2b. So by dividing a^2b over a^3 we'll get the ratio $\dfrac{b}{a}$. Similarly, by dividing ab^2 by a^2b we'll again get $\dfrac{b}{a}$, and for b^3 upon ab^2 we'll get the ratio is $\dfrac{b}{a}$ again.

Now let's consider the second line. Here $2a^2b$ has been doubled from a^2b, and similarly, ab^2 has been doubled to $2ab^2$.

We just need to keep this formula in mind and the rest of the sum automatically follows.

Let's take an example and understand this better.

Say we have to find 12^3.

Here a is 1 and b is 2. So $\dfrac{b}{a}$ is $\dfrac{2}{1} = 2$.

Step 1

We'll cube a, which is $1^3 = 1$. This 1 is the first digit and we'll place it down like this:

1

Step 2

In the subsequent steps, we will simply multiply 1 by $\frac{b}{a}$.
So our first line will look like this:

$$1 \quad 2 \quad 4 \quad 8$$

Step 3

Now we double the two numbers in the middle—2 and
4—to get the second line of the expansion. So 2 becomes
4 and 4 becomes 8. Our sum looks like this now:

$$
\begin{array}{cccc}
1 & 2 & 4 & 8 \\
  & 4 & 8 & \\
\hline
\end{array}
$$

Step 4

This is the final step; all the hard work has already been
done. All we need to do now is simple addition.

- You can see that each column will give us a single digit.
- So in the unit's column we'll bring down 8 and this
  becomes our answer digit for the unit's place.
- Next in the ten's column, we'll add $4 + 8 = 12$. So here
  we'll put 2 down and carry 1 over to the next step.
- In the hundred's column, we'll add $2 + 4 + 1 = 7$, so 7
  is the digit in the hundred's place.
- And finally, we bring down 1.
- So our answer is 1728.

$$
\begin{array}{cccc}
1 & 2 & 4 & 8 \\
  & 4 & 8 & \\
\hline
1 & 7 & 2 & 8 \\
\end{array}
$$

Now let's look at another example, say 13^3.

Here $\frac{b}{a}$ is $\frac{3}{1}$ or 3.

Step 1

As done in the previous example, we'll first cube a, that is $1 - 1^3$ is 1. We now put down 1 like this:

1

Step 2

In the subsequent steps, we will simply multiply 1 by $\frac{b}{a}$.

$\frac{b}{a}$ is nothing but 3 here.

So our first line will look like this:

1   3   9   27

Step 3

Now we double the middle two digits — 3 and 9 — here to get the second line of the expansion. So 3 becomes 6 and 9 becomes 18. Our sum looks like this:

1   3   9   27
     6   18

Step 4

Now it's time to add.

- Remember, each column will give us one single digit.
- So in the unit's column we'll simply bring down 7 from 27 and carry 2 over to the next column. So the unit's place digit becomes 7.
- Now we move over to the ten's column. Here we'll add

9 + 18 + 2 (the carry-over), which gives us 29. We then place 9 down as the ten's digit and carry over 2 to the next column.

- In the hundred's column, we'll add 3 and 6, and the carry-over 2. So that gives us 11. We will then place 1 as the hundred's digit and carry over 1 to the next column.
- Finally, in the thousand's column, we add 1 + 1 = 2.
- Our answer is 2197.

$$
\begin{array}{r}
1\ \ 3\ \ 9\ \ 27 \\
6\ \ 18 \\
\hline
\mathbf{2\ \ 1\ \ 9\ \ 7}
\end{array}
$$

For our next example we'll take 32^3.

Step 1
So 3^3 is 27. This forms the first digit.

27

Step 2
Now $\frac{b}{a} = \frac{2}{3}$. So we'll multiply 27 by $\frac{2}{3}$, which gives us 18—our next digit. We again multiply by $\frac{b}{a}$ in order to get our first line. So our sum now looks like this:

27 18 12 8

Step 3
Now we shall double the middle two numbers.
So 18 becomes 36 and 12 becomes 24. Our sum now looks like this:

$$\begin{array}{rrrr} 27 & 18 & 12 & 8 \\ & 36 & 24 & \\ \hline \end{array}$$

Step 4

Now it's time to add.

- First, we'll put down 8 from the unit's column and now 8 is the answer digit for the unit's place.
- Moving on to the ten's column, we add 12 and 24 and get 36. We'll put 6 down and carry 3 over to the next column.
- Next, moving to the hundred's column, we'll add 18 + 36 + 3 (carried over).
- This gives us 57. We'll now place 7 down and carry over 5 to the next column.
- Finally, we have 27 + 5 = 32 that we put down.
- So our answer is 32768.

$$\begin{array}{rrrr} 27 & 18 & 12 & 8 \\ & 36 & 24 & \\ \hline 3\ 2 & 7 & 6 & 8 \end{array}$$

Let's quickly solve our final example, 38^3.

Step 1

Here $a = 3$ and $b = 8$ and $\dfrac{b}{a} = \dfrac{8}{3}$, so we'll multiply each figure by $\dfrac{b}{a} = \dfrac{8}{3}$.

We find that the cube of 3 is 27; we'll then place it below like this:

27

Step 2

Now in order to get the first line, we will multiply each figure by $\frac{b}{a} = \frac{8}{3}$.

Our first line looks like this:

$$27 \quad 72 \quad 192 \quad 512$$

Step 3

In order to get the second line, we will simply double the middle digits. So 72 becomes 144 and 192 becomes 384. And our sum looks like this:

$$27 \quad 72 \quad 192 \quad 512$$
$$144 \quad 384$$

Step 4

Now it's time to add.

Remember, every column will only give us a single digit.

- So from 512 we'll place 2 below and then carry 51 over to the next column.
- Now we'll add 192 + 384 + 51 = 627. So, placing 7 down in the ten's place, we'll carry 62 over to the next column.
- Now in the hundred's column, we have 72 + 144 + 62 = 278. Here we'll place 8 below and carry 27 over to the next column.
- In the final column, we now have 27 plus 27, which gives us 54.
- So our final answer is 54872.

$$27 \quad 72 \quad 192 \quad 512$$
$$144 \quad 384$$
$$\overline{}$$
$$\mathbf{5\ 4\ 8\ 7\ 2}$$

Maths and Communication

While learning maths, it is important to communicate effectively with your teacher—you should be able to go to your teacher and ask the right questions to understand a topic.

First things first! You should properly understand the name of the topic being taught. You need to know what algebra means or what trigonometry means. 95 per cent of the problems occur when students don't understand what they are studying. So, as a student, please ensure that you understand the meaning of the topic being taught in the class.

Most of the time, students are scared to even approach the teacher. You should always feel free to go to the teacher after class and ask him or her your doubts. If you overcome this fear then you will be able to easily communicate your problems to your teacher.

With the advent of the Internet, student–teacher relationships and internal communication has completely changed. Nowadays, you can even email your questions to your teachers. You can also look up the solution to a problem on the Internet and later discuss it with your teacher or friends.

I'll give you five important tips to be an effective communicator in class.

1. First and foremost, understand the meaning of the topic being taught in class. If everything fails, look it up in a dictionary.
2. Address your questions to the teacher with respect.
3. Use email to contact your teacher.
4. Use the Internet to search for the answer to any maths problem.
5. Every day after class discuss the topic taught and exchange notes with friends.

13

Square Roots

Square Roots of Perfect Squares

THERE ARE TWO methods for calculating square roots using maths sutras! Here we first see how to find the square roots of perfect squares, and in the next chapter, we will look at how to find the square root of any number.

Finding square roots is normally time-consuming and stressful. With the maths sutras, this is going to be child's play. Let's see how.

Here we have made a table with numbers from one to nine with their respective squares, last digits and digit sums.

So we have:

Number	Square	Last Digit	Digit Sum
1	1	1	1
2	4	4	4
3	9	9	9
4	16	6	7

5	25	5	7
6	36	6	9
7	49	9	4
8	64	4	1
9	81	1	9

Looking at the table, we note that:

1. The square of 1 is obviously 1, so the last digit will also be 1, and thus, our digit sum is also 1.
2. Likewise, for 2, its square is 4 and that makes its last digit and digit sum also 4.
3. For 3, the square would be 9, while its last digit and digit sum will be 9 as well.
4. Next, there will be a little change, while the square of 4 is 16; the last digit is 6 as it's a double-digit number, while the digit sum is 7 $(1 + 6 = 7)$.
5. Similarly, the square of 5 is 25. So the last digit is 5 and the digit sum is 7. $2 + 5 = 7$.
6. In the case of 6, the square is 36. So the last digit is 6, while the digit sum is 9 $(3 + 6 = 9)$. Simple, isn't it?
7. The square of 7 is 49. So the last digit here is 9 and the digit sum is 4 $(4 + 9 = 13 = 1 + 3 = 4)$.
8. The square of 8 is 64, which gives us 4 as our last digit, and the digit sum is 1 $(6 + 4 = 10 = 1 + 0 = 1)$. It's that simple.
9. And lastly, the square of 9 is 81. So, as you can see, the last digit is 1, while the digit sum is 9.

Now look at some patterns in the above table.

- The square numbers only have a digit sum of 1, 4, 7 and 9.
- The square numbers only end in 1, 4, 5, 6, 9 and 0.

Based on these points, it's simple to find out if a given number is a perfect square or not, so let's get down to our first example.

Find the square root of 3249.

Step 1
We first put the digits in pairs. Since there are four digits and hence two pairs, we can easily say that the number of digits in our answer will also be two.

Step 2
Now we look at the first two digits — 32 — and see that 32 is more than 25 (5^2) and less than 36 or 6^2. So the answer will be between 50 and 60, because 50^2 is 2500 and 60^2 is 3600. Therefore, the square root of 3249 will be between 50 and 60.

Step 3
In our third step, we focus on the last digit of 3249, which is 9. Any number ending with 3 will end with 9 when it is squared, so the number we are looking for could be 53, but it could also be 57 as 7 squared will also give 9 at the end. So what is our answer — 53 or 57?

Think about it for a moment.

Step 4
Here we can use the digit sums that we know well to find out our answer.

So, if $53^2 = 3249$, then converting it into digit sums we get $(5 + 3)^2 = 8^2 = 64 = 6 + 4 = 10 = 1 + 0 = 1$.

The digit sum of 3249 is $3 + 2 + 4 + 9 = 18 = 1 + 8 = 9$.

Hence, it can't be 53, because the digit sums don't match the digit sum of 3249.

So our answer has to be 57.

We'll now check it: $(5 + 7)^2 = (5 + 7)^2 = 12^2 = (144) = 1 + 4 + 4 = 9$.

Therefore, the digit sum of 3249 must be 9. And when we check, $3 + 2 + 4 + 9 = 9$.

Therefore, our final answer becomes 57.

Let's take another example and solve it.

Let's say we have to find the square root of 2401.

Step 1
As earlier, here too we first put the digits in pairs. Since there are two pairs, we can now say that the number of digits in our answer will also be two.

Step 2
Now, looking at the first two digits — 24, we can see that 24 is greater than 16 (4^2) and less than 25 or 5^2, so the answer will be between 40 and 50.

Step 3
We now look at the end digit in 2401, which is 1.

So the sum could be either 41 or 49 since both squared will give the end digit as 1, because the last digit 41^2 will be $1 \times 1 = 1$, and the last digit of 49^2 will be $9 \times 9 = 1$. I hope this is clear.

Step 4
Now in our final step, we use digit sums to get our final digit. The digit sum of 41^2 is $(4 + 1)^2 = (5)^2 = 25 = 2 + 5 = 7$.

And the digit sum of $49^2 = (4 + 9)^2 = 13^2 = 169 = 1 + 6 + 9 = 16 = 1 + 6 = 7$. So we see that the digit sum of both 41^2 and 49^2 is 7.

We will now try to estimate our answer.

The square of 45 is 2025.

Since 49 is greater than 45, our answer is 49^2, because 2401 is greater than 2025. So the final answer cannot be 41. It has to be 49.

This method of finding square roots is extremely easy and, in fact, can be done mentally.

Let's take another sum and this time we'll try to do it mentally.
Say we need to find the square root of 9604.

Step 1

So as always we first put the digits in pairs. There are two pairs, which means that the answer will have two digits.

Step 2

We take the first pair. The perfect square just less than 96 is 81 (9 squared) and the perfect square just more than 96 is 100 (10 squared).

So, remember, our answer must be between 90 and 100.

Step 3

In the final step, we take 9604, and here the last digit is 4. So the square root can be either 92 or 98, as squaring 92 and 98 will give the last digit as 4.

Let's check 92^2 with digit sums. $(9 + 2)^2 = (11)^2 = 121 = 1 + 2 + 1 = 4$.

After this, let's check $98^2 = (9 + 8)^2 = (17)^2 = (1 + 7)^2 = 8^2$
$= 64 = 6 + 4 = 10 = 1 + 0 = 1$.

Our square root is 98, because the digit sum of 9604 is 1
$(9 + 6 + 0 + 4 = 10 = 1 + 0 = 1)$, which matches with the
digit sum of 98^2. It's that simple!

Now let's take a slightly bigger sum. Say we have to find the
square root of 24964.

Step 1
We first put the digits in pairs. The pairs are 02, 49 and 64.
Please note that there are three pairs now; therefore, we
will have three digits in our final answer.

Step 2
Since 249 lies between $15^2 = 225$ and $16^2 = 256$, the first
two figures must be 15.

Step 3
In our final step, we consider the last figure of our sum, i.e.
4. So the answer can be either 152 or 158. But a wise man
once said that, in life or in maths, you should be doubly
sure! So let us check this using the digit sum.

So the digit sum of 152^2 is $(1 + 5 + 2)^2 = 8^2 = 64 = 6 + 4$
$= 10 = 1 + 0 = 1$.

And the digit sum of 158^2 is $(1 + 5 + 8)^2 = 14^2 = (1 + 4)^2 =$
$5^2 = 25 = 2 + 5 = 7$.

The digit sum of 24964 is $2 + 4 + 9 + 6 + 4 = 25 = 2 + 5 = 7$.
Therefore, the final answer is 158.

Now let's take the final example.

Say we have to find the square root of 32761.

Step 1

We put the digits in pairs as usual. Here we can see that there are three pairs; therefore, we should have three digits in our final answer.

Step 2

Since 327 lies between $18^2 = 324$ and $19^2 = 361$, the first two figures must be 18.

Step 3

As the last figure of our sum is 1, the answer can be either 181 or 189—the last digit of 181^2 as well as 189^2 will be 1. We must find the right answer using the digit sum.

The answer is 181, because the digit sum of 181^2 is $(1 + 8 + 1)^2 = 10^2 = 100 = 1 + 0 + 0 = 1$ and the digit sum of $189^2 = (1 + 8 + 9)^2 = (18)^2 = (1 + 8)^2 = 9^2 = 81 = 8 + 1 = 9$.

The square root of 32761 is 181, because its digit sum is 1.

Square Roots of Imperfect Squares

In this chapter, we will see how to find the square root of any given number. This is an important topic and needs your full attention. But for this, one needs to know the concept of duplex. We already discussed duplexes when we were dealing with squares, so at this stage a review of the topic is highly recommended.

Let's find the square root of 3249.

Step 1

We first divide the number in pairs from right to left. So the first group from the right is 49 and the second group is 32. Since there are two groups, the square root will contain two digits before the decimal.

$$\overline{3\,2}\,\overline{4\,9}$$

Step 2

We take the first group, 32. We then find the perfect square just less than or equal to 32. It is 25 and the square root of 25 is 5. So we bring down 5 as the first square root digit. We then double 5 to make it 10. This 10 is our divisor. Our sum looks like this:

$$10\,\big|\,\overline{3\,2}\,\overline{4\,9}$$
$$5$$

Step 3

We subtract 5^2 (25) from 32. This gives us the remainder 7. We prefix this to 4, making it 74.

$$10\,\big|\,\overline{3\,2\,_74\,9}$$
$$5$$

Step 4

We now divide 74 by 10. This gives us 7 and remainder 4. The remainder is prefixed to 9 making it 49.

$$10\,\big|\,\overline{3\,2\,4\,_49}$$
$$57$$

Step 5

From 49 we subtract the duplex of 7, which is 49. 49 - 49 = 0. And 0 over 10 gives us 0. So our square root becomes 57.

$$\begin{array}{r|l} 10 & \overline{3\,2}\,\overline{{}_4\!9} \\ \hline & 57 \end{array}$$

Basically, there are two steps that you need to keep in mind.

We divide by the double of the first digit and we subtract the duplexes. That's it. It is simple and easy!

Now that we know the rules, let's take another example. Let's find the square root of 529.

Step 1

We divide 529 in pairs from right to left. There will be 2 pairs. The first pair will be 29 and the second one will have only one digit, 5. Since there are two groups, the square root will contain two digits before the decimal. We lay the sum as before.

$$\begin{array}{r|l} & \overline{5}\,\overline{2\,9} \\ \hline & \end{array}$$

Step 2

We then take the first group 5 and find the perfect square just less than or equal to it. The perfect square is 4. We take the square root of 4, which is 2, and put it down as the first digit of our answer. We also double 2 and make it 4. This 4 is the divisor.

$$\begin{array}{r|l} 4 & \overline{5}\,\overline{2\,9} \\ \hline & 2 \end{array}$$

Step 3

We now double 2, this becomes 4. We subtract 4 from 5. This gives us 1. We prefix 1 to 2, making it 12.

$$4\ \overline{\left|\ \begin{array}{ccc} 5, & 2 & 9 \\ \hline 2 & & \end{array}\right.}$$

Step 4

We now divide 12 by 4. This gives us 3, our next answer digit, and remainder 0. We prefix 0 to 9 and this becomes 09.

$$4\ \overline{\left|\ \begin{array}{ccc} 5, & 2 & 9 \\ \hline 23 & & \end{array}\right.}$$

Step 5

From 09 we subtract the duplex of 3. The duplex of 3 is 9. 09 - 9 = 0, and 0 divided by 4 gives us 0. We put our decimal point after two digits. So our answer is 23.0.

$$4\ \overline{\left|\ \begin{array}{ccc} 5, & 2 & 9 \\ \hline 23.0 & & \end{array}\right.}$$

Let's take another example. Say we have to find the square root of 4624.

Step 1

We divide the number 4624 in groups of two from right to left. There will be two groups. This means that our square root will contain two digits before the decimal point.

$$\begin{array}{c|c} & \overline{4\ 6}\ \overline{2\ 4} \\ \hline & \\ \end{array}$$

Step 2

We take 46, and find the perfect square just less than or equal to it, which is 36. The square root of 36 is 6. This becomes our first answer digit. We also double 6 to make it 12. This 12 is our divisor. Our sum looks like this:

$$\begin{array}{c|c} 12 & \overline{4\ 6}\ \underline{2}\ \overline{4} \\ \hline & 6 \\ \end{array}$$

Step 3

From 46 we subtract the square of 6. 46 - 36 = 10. We prefix 10 to 2, making it 102. Our sum looks like this:

$$\begin{array}{c|c} 12 & \overline{4\ 6_{10}}\ \overline{2\ 4} \\ \hline & 6 \\ \end{array}$$

Step 4

After subtraction, we divide. 102 divided by 12 gives us 8 and remainder 6. We put down 8 as our answer digit and prefix 6 to 4 making it 64.

$$\begin{array}{c|c} 12 & \overline{4\ 6_{10}}\ \underline{2}\ 4_6 \\ \hline & 68 \\ \end{array}$$

Step 5

After division, we subtract, 64 minus the duplex of 8. The duplex of 8 is 64. So we have 64 - 64 = 0. 0 divided by 12 gives us 0. So our final answer is 68.0.

$$12 \overline{\smash{\big)}\ 4\ 6_{10}2\ _64}$$
$$68.0$$

Let's do another example together. We have to find the square root of 2209.

Step 1

We divide the number 2209 in groups of two from right to left. There will be two groups. This means that our square root will contain two digits before the decimal point.

$$\overline{2209}$$

Step 2

We first find the perfect square just less than or equal to 22. It is 16. The square root of 16 is 4. This is the first answer digit.

We double 4 to make it 8. 8 is our divisor. Our sum looks like this:

$$8 \overline{\smash{\big)}\ 2209}$$
$$4$$

Step 3

From 22 we subtract the square of 4, which is 16. 22 - 16 = 6. We prefix 6 to 0, this becomes 60.

$$8 \overline{\smash{\big|}\ \overline{2\,2}_6\,\overline{0\,9}}$$
$$\ \underline{4}$$

Step 4

We now divide 60 by 8. This gives us our next answer digit 7 and remainder 4. The remainder 4 is prefixed to 9, making it 49.

$$8 \overline{\smash{\big|}\ \overline{2\,2}_6\,\overline{0_4\,9}}$$
$$\ \underline{47}$$

Step 5

We now subtract the duplex of 7 from 49. The duplex of 7 is 49. 49 - 49 = 0.

0 divided by 8 is 0. So our exact answer is 47.0.

$$8 \overline{\smash{\big|}\ \overline{2\,2}_6\,\overline{0_4\,9}}$$
$$\ \underline{47.0}$$

Let's see another example: $\sqrt{2656}$

Step 1

We divide the number 2656 in groups of two from right to left. There will be two groups. This means that our square root will contain two digits before the decimal point.

$$\overline{\smash{\big|}\ \overline{26}\ \overline{56}}$$

Step 2

We take the first group and find the perfect square just less than or equal to 26. It is 25. The square root of 25 is 5. 5 becomes our first answer digit.

We double 5, which makes it 10. And 10 becomes our divisor. Our sum looks like this:

$$10 \overline{\smash{\big)}\, 2\overline{6}\ 5\overline{6}}$$
$$5$$

Step 3

From 26 we subtract the square of 5. 26 - 25 = 1 .We prefix 1 to 5, making it 15.

$$10 \overline{\smash{\big)}\, 2\overline{6}, 5\overline{6}}$$
$$5$$

Step 4

We divide 15 by 10. It gives us the next answer digit 1 and remainder 5. 5 is prefixed to 6, making it 56. Our sum looks like this:

$$10 \overline{\smash{\big)}\, 2\overline{6}, 5_56}$$
$$51$$

Step 5

Now, from 56, we subtract the duplex of 1, which is also 1. We subtract 56 - 1 = 55. 55 divided by 10 gives us 5 as our answer digit and remainder 5. Our answer is 51.5.

$$10 \overline{\smash{\big)}\, 2\overline{6}, 5_56}$$
$$51.5$$

Be a Maths Achiever: Rising from Your Comfort Zone

Most of the time we do things that come easy to us in maths. We don't take that extra step and do that extra sum, because it might be pushing us too much. We accept that we are meant to be mediocre and never put in that extra effort that will help us stand out.

I remember a story of a champion boxer, who used to practise eighteen hours every day. When someone asked him why he practised so much when he was already a champion, he said, 'My competitor practises seventeen hours every day. Till now we are equals, but this extra hour that I practise gives me the competitive edge, and that's what makes me a champion!'

The same holds true for an achiever in mathematics. You have to work hard on this subject in order to stand out as an achiever. You have to overcome your fear of maths by coming out of your comfort zone. That extra step you take or the extra hour you put in will make you the achiever!

Let's see how one becomes an achiever in maths:
- Ask your maths teacher smart questions and fully understand the topic taught.
- Solve all the problems from all the sources and books available.
- Solve the previous year's model test papers.
- Have a full grasp of the concept taught in class.
- Help other students who are struggling with mathematics.
- Check and recheck the paper before submitting.
- Research the Internet about topics on maths and participate in intelligent discussions.

14

Cube Roots

IN 1977, at Southern Methodist University, Shakuntala Devi was asked to give the twenty-third root of a 201-digit number; she answered in fifty seconds. Her answer — 546,372,891 — was confirmed by calculations done at the US Bureau of Standards by the UNIVAC 1101 computer, for which a special program had to be written to perform such a large calculation.

Ms Shakuntala Devi
(1929–2013)

Shakuntala Devi was India's prodigy and was popularly known as the 'human computer'. In her academic shows, she used to dazzle the audience with her mathematical feats, one of which was finding the cube root of a six-digit number in a matter of seconds. Here we see one of the secrets of her feat to finding the cube roots of six-digit numbers. This method, when extended, gives the secret to calculate even the twenty-third root of a number.

This table shows us the cubes of numbers from 1 to 9.

Number	Cubes
1	1
2	8
3	27
4	64
5	125
6	216
7	343
8	512
9	729

Let us see some patterns in this table.

If the cube ends in a 1, the number or the cube root of the same number also ends in a 1. Similarly, if the cube ends in a 4, 5, 6 and 9, its cube root always ends in a 4, 5, 6 and 9. Let's see:

Number	Cubes
1 ←——— 1
2	8
3	27
4 ←——— 64
5 ←——— 125

| 6 ◄——————216 |
| 7   343 |
| 8   512 |
| 9 ◄——————729 |

Now we also see that if the cube ends in an 8, the cube root ends in a 2. If the cube root ends in a 2, the cube ends in an 8.

Number	Cubes
1	1
2 ◄———————8	
3	27
4	64
5	125
6	216
7	343
8 ◄———————512	
9	729

If the cube ends in a 7, its cube root will end in a 3 and if the cube root ends in a 3, its cube ends in a 7.

Number	Cubes
1	1
2	8
3 ◄———————27	
4	64
5	125
6	216
7 ◄———————343	
8	512
9	729

The Method

Suppose we have to find the cube root of a perfect cube 3375.

Step 1

We divide the perfect cube to groups of three, because this is a cube, starting from right to left. Here, in 3375, the first group of three is 375 and the second group only has 3. Now since there are only two groups, our cube root or the answer will have only two digits.

$$\overline{3}\,\overline{375}$$

Step 2

From the last digit of the given cube we find the last digit of the cube root.

Here the last digit is 5, so the last digit of the cube root will also be 5. This is very easy!

Step 3

To get the first digit of the cube root, we simply find the perfect cube less than or equal to 3.

The perfect cube less than or equal to 3 is 1. The cube root of 1 is also 1.

So our answer, the cube root of 3375, is 15.

$$\overline{3}\,\overline{375} = 15^3$$

Let's take another example. Say, we have to find the cube root of the perfect cube 328509.

Step 1

We divide the number in groups of three from right to left. Since there are only two groups, the cube root will have only two digits.

$$\overline{328}\,\overline{509}$$

Step 2

From the last digit of the given cube we find the last digit of the cube root, which is 9. This is very easy!

Step 3

To find the first digit of the cube root we have to find the perfect cube less than or equal to 328.

The perfect cube less than or equal to 328 is 216. The cube root of 216 is 6, and this is the first digit of the cube root. So our cube root becomes 69.

$$\overline{328}\,\overline{509} = 69^3$$

We must remember that, in order to find the cube root, the given number must be a perfect cube. Now you must be wondering how we are to know if a given number is a perfect cube or not.

Well, the solution lies in digit sums. Finding the sum of all the digits reveals a pattern. Let's see how:

Number	Cubes	Digit Sum
1	1	1
2	8	8
3	27	9
4	64	1
5	125	8

6	216	9
7	343	1
8	512	8
9	729	9

The sum of the cube 1 is 1, 8 is 8, etc., as they are all single figures.

The sum of the digits of 27 is $2 + 7 = 9$.

Similarly, the sum of the digits of 64 is $6 + 4 = 10$. Now we have $1 + 0 = 1$.

The digit sum of 125 is $1 + 2 + 5 = 8$. And the digit sum of 216 is $2 + 1 + 6 = 9$.

So we are seeing a pattern of 1–8–9.

Therefore, the digit sum of the cube would be 1, 8 or 9. So any perfect cube will have a digit sum of 1, 8 or 9. So that's how you can identify whether a given number is a perfect cube or not.

But you must remember that a number that has a digit sum of 1, 8 or 9 does not necessarily have to be a cube. Even if it is a cube, it may not be a perfect cube.

Let's take another perfect cube, 175616, and find its cube root.

Step 1
We divide the number in groups of three from right to left. Since there are only two groups, the cube root will have only two digits.

$$\overline{175}\,\overline{616}$$

Step 2

From the last digit of the given cube we find the last digit of the cube root, which is 6.

Step 3

To find the first digit of the cube root, we take the first group, which is 175. We find the perfect cube just less than equal to 175. This is 125. Then we find the cube root of 125, which is 5—this is the first digit of the cube root.
So our cube root is 56!

There are a few things that you must remember:

1. This method works for perfect cubes only.
2. The method for working out the cube root of any larger number is very exhaustive, so we restrict our finding of cube roots to numbers that have six digits or fewer.

So let's find the cube root of another perfect cube, 373248.

Step 1

We divide the number in groups of three from right to left. Since there are only two groups, the cube root will have only two digits.

$$\overline{373}\,\overline{248}$$

Step 2

From the last digit of the given cube we find the last digit of the cube root, which is 2.

Step 3

Now to find the first digit of the cube root, we take the first group, which is 373.

We find the perfect cube just less than or equal to 373, which is 343.

The cube root of 343 is 7—this is our first digit.

So the cube root of 373248 is 72.

$$\overline{37}\overline{3248}=72^3$$

Relaxation and Concentration Exercise

How do you feel just before a maths exam? Tensed? Nervous? Do you also feel sick? I know it because, I confess, I used to feel sick too, especially before a maths exam.

I am going to share another deep secret with you. If you are stressed before an exam, it directly affects your exam score. Your nervousness may even cause you to forget the maths formulas, and this will result in low marks.

Today, we will see how to manage stress during exams and how to succeed in your maths paper. I am going to teach you a breathing exercise that will help reduce your stress levels. This exercise is called Anuloma Viloma or Alternate Nostril Breathing.

The exercise of the Anuloma Viloma makes both sides of the brain, the left side which is responsible for logical thinking and the right side which is responsible for creative thinking, to function properly. This will lead to a balance between a person's creative and logical thinking. Yogis consider this to be the best technique to calm the mind and the nervous system.

So let's practise the breathing technique:

Inhale through your left nostril, closing the right with your thumb, to the count of four.

Hold your breath, closing both nostrils, to the count of ten.

Exhale through your right nostril, closing the left with the ring and little fingers, to the count of eight.

Inhale through the right nostril, keeping the left nostril closed with the ring and little fingers, to the count of four.

Hold your breath, closing both nostrils, to the count of ten.

Exhale through the left nostril, keeping the right closed with the thumb, to the count of eight.

Stay Stress-free

This yogic technique is very potent and makes you stress-free. It also improves your concentration. Do this just before your exam.

I am also sharing three techniques with you to improve concentration and reduce stress and worry.

1. **Listen to music:** You feel good too, which increases your confidence and marks in the maths paper.
2. **Call up a friend:** You can call a friend or a cousin and talk for some time. This is considered a stress buster.
3. **Listen to a joke:** Laughter is a good therapy for stress! Nothing can beat the power of a good laugh.

15

Raising to Fourth and Higher Powers

IN THIS CHAPTER, we will be seeing how to raise any number to the fourth power.

Raising a Number to the Fourth Power

The method to raise a number to the fourth power is very much like the method of cubes.

So let's see what $(a + b)$ is when raised to the power 4.

$$(a + b)^4 = a^4 + 4a^3b + 6a^2b^2 + 4ab^3 + b^4$$

$$
\begin{array}{r}
(a+b)^4 = a^4 + a^3b + a^2b^2 + ab^3 + b^4 \\
3a^3b + 5a^2b^2 + 3ab^3 \\
\hline
a^4 + 4a^3b + 6a^2b^2 + 4ab^3 + b^4
\end{array}
$$

In this formula, consider a to be in the ten's place and b in the unit's place. Here you will spot a ratio between a^4 and a^3b. Divide a^3b by a^4 and you will get the ratio $\dfrac{b}{a}$. Similarly, divide a^2b^2 by a^3b and you will get $\dfrac{b}{a}$ again. Follow the same for ab^3 and b^4. You will get the same ratio $\dfrac{b}{a}$.

Now consider the second line. Here a^3b and ab^3 has been tripled to $3a^3b$ and $3ab^3$. a^2b^2 has been multiplied five times to make it $5a^2b^2$.

Now that we have understood this concept, let's solve some examples.

Say we have to find 12^4. Here a is 1 and b is 2. So $\dfrac{b}{a}$ is $\dfrac{2}{1} = 2$.

Step 1

We raise a to the power of 4, which means $1^4 = 1$. This 1 is the first digit. We put it down like this:

1

Step 2

We then multiply the subsequent digits by 2 as $\dfrac{b}{a}$ is $\dfrac{2}{1} = 2$. So our first line looks like this:

1 2 4 8 16

Step 3

We will now do the next step for the second line.

2 gets multiplied by 3 to become 6.

4 gets multiplied by 5 to become 20.

And 8 gets multiplied by 3 to become 24.

Our sum looks like this now:

1 2 4 8 16
  6 20 24

Step 4

In our final step, we start adding from right to left. Remember, each column will give us a single digit.

- From 16 we put down 6 in the unit's place and carry over 1 to the next step.
- In the ten's place, we add $8 + 24 + 1$ (carried over) = 33. We put down 3 in the ten's place and carry over 3 to the next step.
- In the hundred's column, we have $4 + 20 + 3$, this gives us 27. We put down 7 and carry over 2 to the next step.
- In the thousand's place, we have $2 + 6 + 2$ (carried over) = 10. We put down 0 and carry 1 to the next step.
- In the final step, we have $1 + 1 = 2$.
- So our final answer is 20736.

Our sum on completion looks like this:

$$
\begin{array}{r}
1\ 2\ 4\ 8\ 16 \\
6\ 20\ 24 \\
\hline
2\ 0\ 7\ 3\ 6
\end{array}
$$

Now that we have learnt this, let us apply the rules to calculate 13^4.

Here clearly $\dfrac{b}{a}$ is $\dfrac{3}{1} = 3$.

Step 1

We take a, which is 1, and raise it to power of 4. $1^4 = 1$. Our sum looks like this:

1

Step 2

We then multiply the subsequent digits by 3.
Our first line looks like this now:

$$1 \quad 3 \quad 9 \quad 27 \quad 81$$

Step 3

Now to get the second line, we multiply 3 by 3, 9 by 5 and 27 by 3, just like we saw in the expansion of $(a + b)^4$. So our second line becomes:

$$
\begin{array}{ccccc}
1 & 3 & 9 & 27 & 81 \\
  &   & 9 & 45 & 81 \\
\hline
\end{array}
$$

Step 4

We now add from right to left.

- We put down 1 of 81 and carry over 8 to the next column.
- $27 + 81 + 8$ (carried over) $= 116$. We put down 6 and carry 11 to the next column.
- $9 + 45 + 11 = 65$. We now put down 5 and carry over 6 to the next column.
- $3 + 9 + 6 = 18$. We put down 8 and carry over 1 to the next column.
- $1 + 1 = 2$.
- Our complete answer is 28561.

$$
\begin{array}{ccccc}
1 & 3 & 9 & 27 & 81 \\
  &   & 9 & 45 & 81 \\
\hline
\multicolumn{5}{c}{\textbf{28561}}
\end{array}
$$

Let's now take another sum. Say we have 32^4. Here $\dfrac{b}{a}$ is $\dfrac{2}{3}$, so our ratio is $\dfrac{2}{3}$.

Step 1

We take a, which is 3 and raise it to the power of 4.

$a^4 = 3^4 = 81$

Our sum looks like this:

81

Step 2

Our first line looks something like this:

81 54 36 24 16

Step 3

Now to get the second line, we multiply 54 by 3, 36 by 5, and 24 by 3, just like we saw in the expansion of $(a + b)^4$. So our second line becomes:

81 54 36 24 16

162 180 72

Step 4

In our final step, we just add right to left.

- From right to the left, we put down 6 and carry 1 to the next column.
- $24 + 72 + 1$ (carried over) = 97. We put down 7 and carry 9 to the next column.
- $36 + 180 + 9 = 225$. We put down 5 and carry 22 to the next column.
- $54 + 162 + 22 = 238$. We put down 8 and carry over 23 to the next column.
- Finally, we have $81 + 23 = 104$.

- Our answer is 1048576.

$$81 \quad 54 \quad 36 \quad 24 \quad 16$$
$$\underline{162 \quad 180 \quad 72}$$
$$\mathbf{1048576}$$

We can apply the same principles and raise the number to fifth and sixth powers as well. The method is the same, and I am sure you would like to try it out for yourself.

To get you started I am giving you the expansion of $(a + b)^5$ here.

$(a + b)^5 = a^5 + 5a^4b + 10a^3b^2 + 10a^{2b3} + 5ab^4 + b^5$

Go ahead now and find 12^5.

Seven Important Maths Test Tips

For most students the night before a mathematics exam is spent in tension. There is hardly any time to rest between revising notes, learning formulas, solving all the practice problems from the textbook, all in one night, is there?

Today, we'll discuss some important tips for taking maths tests.

1. Repetition is important in maths. You learn to do the sums correctly after a lot of practice. So keep practising, but don't try to solve the sums mechanically—learn the steps diligently and then solve the sums.

2. Each topic has problems that have various difficulty ranges. Ensure that you do the easier ones first before moving on to the harder ones. Don't hesitate to ask your teacher for help if you are stuck.

3. Prepare a sheet with all the formulas that you need, just like flash cards, and then it becomes easy to remember it.

4. During a test, first write all the formulas you remember on the back of the paper. This helps as there is a high possibility that you might forget them later.

5. Read all the instructions carefully before attempting the problems.

6. Please show all the working notes for the problem. Even if you know that it's the wrong answer, don't erase your work. Let your workings be, since you may get marks for the steps that are correct.

7. Check your answer paper after you have finished. If you have time, redo the problems on a separate sheet of paper and see if you get the same answers the second time too. Look out for careless mistakes—make sure the decimal is in the right place, read the directions correctly, copy the numbers attentively, put a negative sign if it is needed, etc.

If you keep these seven tips in mind, I am sure any test will be a smooth ride. Not only will you pass the test with flying colours but it will also build your confidence in maths. Trust me on this!

16

Algebraic Calculations

GET READY TO ignite your brain cells with some amazing maths sutra–style algebraic calculations. Let's launch into our first example.

Let's take $(x + 2)(x + 3) + (x - 6)(x + 5)$.

No need to write this down. Let's learn how to do some mental calculations. Here we can mentally note down the answer.

Step 1
We solve this using the vertically and crosswise method.
First, we multiply $(x + 2)$ and $(x + 3)$.
We can either implement the method left to right or from right to left.
$(x + 2)$
$(x + 3)$

Going from left to right, we multiply vertically. So we have x times x, which gives us x^2. So this is the first part of step 1.

Similarly, we do the same for $(x - 6)$ and $(x + 5)$, that is, we multiply the x's vertically.

So we have x^2 here as well. Adding both these x^2 terms we get $2x^2$.

Step 2

Now going with the second step, we multiply crosswise.
So we have
$(x + 2)$
$(x + 3)$

Here we cross-multiply and get three times x plus x times 2. We add $3x + 2x = 5x$.
Now we have:
$(x - 6)$
$(x + 5)$

Here we cross-multiply and get five times x plus x times -6. We add $5x$ and $-6x$ and get $-x$.
Combining the two answers, we get
$5x - x = 4x$

Step 3

In our final step, we'll get +6 by multiplying 3 and 2 together and -30 (by multiplying -6 and 5 together), which gives us -24 as the next part of our answer.

So our answer is $(x + 2)(x + 3) + (x - 6)(x + 5) = 2x^2 + 4x - 24$.
Wasn't this incredibly simple?

Let's take another example: $(2x - 3)(x + 5) + (x - 1)(x + 3)$.

Step 1
We solve this using the vertically and crosswise method.
So first, we multiply vertically the terms which have x.
$(2x - 3)$
$(x + 5)$

So we multiply $2x$ and x vertically and we get $2x^2$.
Similarly, we multiply vertically the terms which have x
here:
$(x - 1)$
$(x + 3)$

So x times x gives us x^2.
Adding, $2x^2 + x^2 = 3x^2$.

Step 2
Now to get the values of the middle term we multiply
crosswise here:
$(2x - 3)$
$(x + 5)$

And get $((5 \times 2x) - (3 \times x)) = 10x - 3x = 7x$.
Similarly, we multiply crosswise here too mentally:
$(x - 1)$
$(x + 3)$

And get $(3 \times x) + (-1 \times x) = 3x - x = 2x$.
Combining $7x$ and $2x$ gives us $9x$.

Step 3

Finally, we multiply -3 and 5 vertically to get -15.

We multiply -1 and 3 vertically to get -3.

Combining these two together we get -15 and -3, which equals -18.

So our final answer after combining steps 1, 2 and 3 is $3x^2 + 9x - 18$.

We just did all the calculations one by one mentally and here's our answer! Now let's look at some more examples and, this time too, let's try to do the calculations mentally!

Say we have $(2x - 3)^2 - (3x - 2)(x + 5)$.

Step 1

We solve this using the vertically and crosswise method.

First, we multiply vertically the terms that have x. So we have

$(2x - 3)$

$(2x - 3)$

$2x \times 2x = 4x^2$

And

$(3x - 2)$

$(x + 5)$

Here, multiplying vertically we get $3x \times x = 3x^2$.

So mentally, we do it like this: $4x^2 - 3x^2 = x^2$.

This is the first part of our answer.

Step 2

Next, we multiply crosswise.

So, as the first part, we have

$(2x - 3)$

$(2x - 3)$

So we get from the first product $-2x \times 3$ and $-2x \times 3 = -6x + -6x = -12x$.

And from the second product, we get $5 \times (3x) - 2 \times x$.

$(3x - 2)$

$(x + 5)$

So we have $15x - 2x = 13x$.

So combining the first part with the second part we have $-12x + -13x = -25x$.

Step 3

And in the final step, we add the independent terms, $+9$ and $+10$ and get 19.

So our final answer is $(2x - 3)^2 - (3x - 2)(x + 5) = x^2 - 25x + 19$.

For the next sum, let's take $(5x - 2)^2 - (2x + 1)^2$.

Step 1

Using the same method, we first multiply vertically the terms that have x.

So we have

$(5x + 2)$

$(5x + 2)$

So $5x \times 5x = 25x^2$.

And we have

$(2x + 1)$

$(2x+1)$

So $2x \times 2x = 4x^2$.

Now we'll take the terms containing x^2: $25x^2$ (from the first product) - $4x^2$ (from the second product) = $21x^2$.

Step 2

Now in this step, we multiply crosswise. So we have

$(5x + 2)$

$(5x + 2)$

Here we multiply $5x \times 2$ twice = $10x + 10x = 20x$.

Again we have

$(2x + 1)$

$(2x + 1)$

Here we cross-multiply $2x \times 1$, both ways, and get $2x$ both times.

We add $2x$ and $2x$ and get $4x$.

Altogether, this becomes $20x - 4x = 16x$.

Step 3

And in our final step, we have the independent digits $2^2 - 1^2$ = $4 - 1 = 3$.

So combining all the parts, we get

$(5x + 2)^2 - (2x + 1)^2 = 21x^2 + 16x + 3$.

Now let's solve our next example:

$(2x + 3y + 4)(x - 3y + 5)$

Now these kind of sums can be done easily using the vertically and crosswise method of multiplication. So what are we waiting for? Let's do a quick recap of this method.

3-Digit Vertically & Crosswise Pattern

There you have it. Now with the pattern shown, let's straightaway go to our example and apply it.

Step 1
Here we'll solve the sum right to left, so we first multiply vertically 5 and 4, and that gives us 20.

$$2x + 3y + 4$$
$$\underline{x - 3y + 5}$$
$$+20$$

Step 2
Next, we'll solve it crosswise: $(5 \times 3y) + (4 \times 3y)$.
So we get $15y - 12y = 3y$.

$$2x + 3y + 4$$
$$\underline{x - 3y + 5}$$
$$+3y+20$$

Step 3

Now comes the third step—the star multiplication step.

So we multiply $(5 \times 2x) + (4 \times x) + (3y \times -3y)$.

And this becomes $10x + 4x - 9y^2 = 14x - 9y^2$.

$$2x + 3y + 4$$
$$\underline{x - 3y + 5}$$
$$14x-9y^2+3y+20$$

Step 4

Now we'll solve it crosswise again as per the pattern.

So we have $(2x \times -3y) + (x \times 3y)$.

So this becomes $-6xy + 3xy = -3xy$.

$$2x + 3y + 4$$
$$\underline{x - 3y + 5}$$
$$-3xy+14x-9y^2+3y+20$$

Step 5

In our final step, we will follow the vertical pattern.

So when we multiply $2x$ and x, it gives us $2x^2$.

$$2x + 3y + 4$$
$$\underline{x - 3y + 5}$$
$$2x^2-3xy+14x-9y^2+3y+20$$

I assure you that with a little bit of practice you will be able to solve even bigger sums.

Visualizing Success in Maths

I am going to share with you some secrets to score 100 per cent in mathematics.

To excel in maths you must really understand where you want to stand and you must visualize yourself achieving that success.

Draw pictures of your success in maths and post it on your wall. Yes, you can also post them on your Facebook wall to declare your dreams to all your friends. You can visualize your yourself as a 100 per cent scorer in maths, and slowly, this picture will inspire you to get that 100 per cent.

Let's see some of the ways to boost your self-image.

1. Make a picture of you accepting an award from your principal for scoring that 100 per cent in maths. Yes, it's

easy! If you want you can even write an acceptance speech. Yes, sometimes getting that 60 per cent is also a remarkable achievement for a slow learner. So you could move slowly from 40 to 60 per cent, from 60 to 80 per cent and finally, from 80 to 100 per cent.

Put this picture everywhere around you — on your wall, on your desktop or even in your wallet! You should be passionate about maths.

2. Make a hand-drawn Certificate of Merit or a report card. Write down your estimated score for maths and, when you see this every day, you will get inspired.

3. The more you can articulate success and make constant reminders (by posting pictures), the more deeply you plant the seeds of success in your mind, and the greater the chance for you to experience it too. As James Allen said, 'As a man thinketh in his heart, so he is.'

17

Simultaneous Equations

IN LIFE OR IN MATHS, building strong equations are most important and understanding the relationship between any two variables is the key. Unfortunately, for most students the term 'equations' means complications and endless steps, but what if I tell you that you can kill two birds with one stone? With the help of the maths sutras, not only can we do away with all those complications but it also makes the process super fast.

We'll learn how to mentally solve simultaneous equations. Simultaneous equations are those equations that have two unknown variables x and y. Here we have to find the values of both x and y.

Simultaneous Equations

To play a game, you first need to learn the rules. There is a general formula to solve simultaneous equations and all we have to do is apply this formula to get the values of x and y.

Say the equations are in the form of

$ax + by = p$

$cx + dy = q$

To find the value of x and y, we'll apply the following formula

$$x = \frac{bq - pd}{bc - ad}$$

$$y = \frac{pc - aq}{bc - ad}$$

Let's understand this better by using an example.

$2x + 3y = 8$
$4x + 5y = 14$

Step 1
We'll first find the value of x.

$$x = \frac{bq - pd}{bc - ad}$$

So we'll write $x = \frac{(3 \times 14) - (5 \times 8)}{(3 \times 4) - (2 \times 5)}$.

Now $x = \frac{42 - 40}{12 - 10} = \frac{2}{2} = 1$.

Step 2
Now we find the value of y.

$$y = \frac{pc - aq}{bc - ad}$$

$$y = \frac{(8 \times 4) - (2 \times 14)}{(3 \times 4) - (2 \times 5)}$$

$$y = \frac{32-28}{12-10} = \frac{4}{2} = 2$$

So here the value of $x = 1$ and $y = 2$.

The key is to visualize the pattern in your mind. So to find the value of x, we'll use the following pattern.

$$2x + 3y = 8$$
$$4x + 5y = 14$$

Here we have the numerator as shown by the arrows. All we need to do is follow the arrows and solve crosswise as shown in the example.

The numerator becomes $(3 \times 14) - (5 \times 8)$.

And for the denominator, we'll visualize as shown below:

$$2x + 3y = 8$$
$$4x + 5y = 14$$

So here we have $(3 \times 4) - (2 \times 5)$ as our denominator. Therefore,

$$x = \frac{(3 \times 14) - (5 \times 8)}{(3 \times 4) - (2 \times 5)}$$

I hope the visual pattern of crosswise multiplication is clear.

Let's take another example and apply the visual pattern. Remember, practice makes a man perfect!

Our sum is　$3x + 5y = 19$
$$2x + 3y = 12$$

Let's apply the pattern to this.

Here we'll first find the value of x. Just follow the pattern the arrows are showing. Look at them as visual cues.

$$3x + 5y = 19$$
$$2x + 3y = 12$$

The numerator becomes $(5 \times 12) - (3 \times 19) = 60 - 57 = 3$.

$$3x + 5y = 19$$
$$2x + 3y = 12$$

Following the pattern our denominator becomes $(5 \times 2) - (3 \times 3) = 10 - 9 = 1$.

So the value of x becomes $x = \dfrac{3}{1} = 3$.

We now need to find the value of y.

$$3x + 5y = 19$$
$$2x + 3y = 12$$

Now the numerator becomes $(19 \times 2) - (3 \times 12) = 38 - 36 = 2$. And the denominator will be:

$$3x + 5y = 19$$
$$2x + 3y = 12$$

The denominator here is $(5 \times 2) - (3 \times 3) = 10 - 9 = 1$.

So y becomes

$$y = \frac{(19 \times 2) - (3 \times 12)}{(5 \times 2) - (3 \times 3)}$$

$$y = \frac{38\text{-}36}{10\text{-}9} = \frac{2}{1} = 2$$

As we keep solving more and more examples, this method will become clearer to you.

Let's take another example to understand this method better.

Say we have $\begin{aligned} 2x + 3y &= 14 \\ 5x + 7y &= 33 \end{aligned}$

Now let's solve this mentally! See if you can get the values of x and y.

Just close your eyes and visualize the pattern.

Let's see the value of x first. We see our numerator, which is $(3 \times 33 - 7 \times 14)$.

$$\begin{aligned} 2x + 3y &= 14 \\ 5x + 7y &= 33 \end{aligned}$$

And our denominator, which is $(3 \times 5 - 2 \times 7)$.

$$\begin{aligned} 2x + 3y &= 14 \\ 5x + 7y &= 33 \end{aligned}$$

So x becomes:

$$x = \frac{(3 \times 33)\text{-}(7 \times 14)}{(3 \times 5)\text{-}(2 \times 7)}$$

$$y = \frac{(14 \times 5)\text{-}(2 \times 33)}{(3 \times 5)\text{-}(2 \times 7)}$$

We'll now find the value of *y*.

$$2x + 3y = 14$$
$$5x + 7y = 33$$

Here our numerator is (14 x 5 - 2 x 33).
And our denominator is (3 x 5 - 2 x 7).

$$2x + 3y = 14$$
$$5x + 7y = 33$$

So *y* becomes

$$y = \frac{(14 \times 5)-(2 \times 33)}{(3 \times 5)-(2 \times 7)}$$

$$y = \frac{70-66}{15-25} = \frac{4}{1} = 4$$

I hope that by now you have understood how to solve simultaneous equations completely in your head!

Now let's see a special type in simultaneous equations. Say we have:

$3x + 2y = 6$
$9x + 5y = 18$

Notice that the coefficients of *x* are 3 and 9. So they are in the ratio of 1:3 and the ratio of the independent terms on the right-hand side—6 and 18—is also 1:3.

So we have 3:9 = 6:18.

In such a case, we need to apply the maths sutra—If One Is in Ratio, the Other One Is Zero. Now this tells us that if *x* is in ratio to the independent variables, the other variable, y, is 0. So here $y = 0$.

If $y = 0$, then we can easily find the value of x by substituting y's value in any one of the equations.

So we get

$3x + 0 = 6$

$x = \dfrac{6}{3} = 2$

Now here $x = 2$ and $y = 0$.

So let's take another example of a similar nature.

$4x - y = 20$
$x + 5y = 5$

Here the coefficients of x are 4 and 1; they are in the ratio of 4:1. The ratio of the independent terms on the right-hand side—20 and 5—are also in the ratio of 4:1.

So we have 4:1 = 20:5.

This is the clue we needed, so now we'll apply the maths sutra, if one is in ratio, the other one is zero. So here again if x is in ratio to the independent terms, y is 0.

We can easily find out the value of x from any one of the equations. So here we have:

$x + 5y = 5$
$x + 0 = 5$
$x = 5$

Now $x = 5$ and $y = 0$.

In this chapter, we learnt how to solve simultaneous equations using a brand new method. Solving the unknown variables in the *Maths Sutra* way not only makes calculations easier, but also makes us super-efficient in our calculations.

Success in Maths

Let me indulge in some nostalgia. When I look back and reflect on my schooldays, I realize there were so many things I could have done better, especially in maths. I thought for some time and made a list of things that could have helped me increase my scores in maths.

Let's discuss the factors which take away success and contribute to failure in maths:

- Firstly, we are scared of maths itself.
- We hardly focus on maths. We give this subject the last priority.
- We are not motivated enough to do sums in maths.

- Some maths teachers do not know how to teach properly. This causes a lack of interest in maths.
- We have no time to do our maths homework, because we use it to surf the Net or play games.

If you want to increase your grades to A-stars or a 100 per cent in maths, you have to do certain things. Let's look at some qualities that make a person successful in maths:

- You have a positive mental attitude towards maths. You are comfortable with numbers and are not terrified of it.
- You are a good listener and you follow your teacher's methods in a disciplined way. You consider your teacher to be your mentor!
- You are motivated by maths and you take an interest in the subject. Vedic maths will help to sustain your interest in the subject, since it makes the dull and boring maths seem fun!
- You regularly follow the maths class so much that you are sometimes ahead of it.
- You make a list of practical/realistic goals in maths and know how much time to give to each topic. You work on your goals regularly and achieve them too.
- Simply remember these guidelines to become a maths champion yourself!

18

Quadratic Equations

MATHEMATICAL EQUATIONS HAVE several usages in everyday life, and in this chapter, we'll learn how to solve quadratic equations using Vedic maths. So if you dream of becoming an engineer or any other successful professional, then you will need to crack competitive exams. This is where the world's fastest mental maths system helps you. Let's take a look at what quadratic equations are.

Quadratic equations are of the form:

$$ax^2 + bx + c = 0$$

Quadratic means 'two'. So here the unknown variable x will have two values.

Keep in mind that, in a quadratic equation, there's a simple relationship between the differential, which is $2ax + b$, and the discriminant, which is $b^2 - 4ac$, and the differential is equal to the square root of the discriminant. So in a quadratic expression $ax^2 + bx + c$, the first differential will be $2ax + b$, while the discriminant will be $b^2 - 4ac$.

Here we have the formula for solving the quadratic equation as:

$$x = \frac{-b \pm \sqrt{b^2 - 4ac}}{2a}$$

$$2ax = -b \pm \sqrt{b^2 - 4ac}$$

$$2ax + b = \pm \sqrt{b^2 - 4ac}$$

> So we transpose and adjust to get the relationship between the differential and the discriminant.

Now we will use this relationship to find out the two values of x.

Here we have our first sum:

$7x^2 - 5x - 2 = 0$

Here $a = 7$, $b = -5$ and $c = -2$.

So we apply the formula $2ax + b = \pm \sqrt{b^2 - 4ac}$

Here, the differential is the square root of the discriminant. So we'll get

$$14x - 5 = \pm \sqrt{25 - (4 \times 7 \times -2)}$$

$$14x - 5 = \pm \sqrt{81}$$

$$14x - 5 = \pm 9$$

Now $x = \dfrac{9+5}{14} = 1$ and $x = \dfrac{-9+5}{14} = -\dfrac{2}{7}$ and there you have it, our final answer.

Let's take another example, $6x^2 + 5x - 3 = 0$.

Here $a = 6$, $b = -5$ and $c = -3$.

So we have the formula $2ax + b = \pm\sqrt{b^2 - 4ac}$.

Applying the same principles here, we'll get the differential $2ax + b$ as $12x + 5$, which is equal to the square root of the discriminant. So what we'll get is:

$$6x^2 + 5x - 3 = 0$$
$$12x + 5 = \pm\sqrt{25 - (4 \times 6 \times -3)}$$
$$12x + 5 = \pm\sqrt{97}$$
$$x = \frac{-5 \pm \sqrt{97}}{12}$$

All done in just three steps! Let's now take another example and understand this better.

$$2x^2 - 5x + 2 = 0$$
Here $a = 2$, $b = -5$ and $c = 2$.

So we apply the formula, $2ax + b = \pm\sqrt{b^2 - 4ac}$.

Applying the same principles, we'll get the differential $2ax + b$ as $4x - 5$ and this is equal to the square root of discriminant. So now we get:

$$2x^2 - 5x + 2 = 0$$
$$4x - 5 = \pm\sqrt{25 - (4 \times 2 \times 2)}$$
$$4x - 5 = \pm\sqrt{9}$$
$$x = \frac{5 \pm \sqrt{9}}{4}$$

Done. Moving on, let's look at some special cases of quadratic equations.

Special Case of Quadratic Equations — Reciprocals

Case A

Say we have $x + \dfrac{1}{x} = \dfrac{17}{4}$.

Solving this by the conventional method would take a lot of time and steps. But we can easily solve it using Vedic maths and a little bit of observation. We can see that the left-hand side is the sum of two reciprocals, so we'll simply split the $\dfrac{17}{4}$ on the right-hand side into $4 + \dfrac{1}{4}$.

And what we get is $x = 4$ or $\dfrac{1}{4}$. This is nothing but simple observation.

Let's take another similar example, $x + \dfrac{1}{x} = \dfrac{26}{5}$.

Just like we did in the last sum, here too we'll split the right-hand side into $5 + \dfrac{1}{5}$.

And therefore, we get $x = 5$ or $\dfrac{1}{5}$.

Let's take another sum $\dfrac{x}{x+1} + \dfrac{x+1}{x} = \dfrac{82}{9}$.

Let's make it real quick! Here again we can see that the left-hand side is the sum of reciprocals. Therefore, let's split the right-hand side into $9 + \dfrac{1}{9}$.

And that gives us

$$\frac{x}{x+1} = 9$$

$$x = 9x + 9$$

$$x - 9x = 9$$

$$-8x = 9$$

$$x = \frac{-9}{8}$$

Or

$$\frac{x}{x+1} = \frac{1}{9}$$

$$9x = x + 1$$

$$9x - x = 1$$

$$8x = 1$$

$$x = \frac{1}{8}$$

So $x = \dfrac{-9}{8}$ or $x = \dfrac{1}{8}$.

Now let's quickly solve another example of reciprocals,

$$\frac{x+1}{x+2} + \frac{x+2}{x+1} = \frac{37}{6}$$

Here, as earlier, we'll first split the right-hand side into reciprocals 6 and $\dfrac{1}{6}$.

$$\frac{x+1}{x+2} + \frac{x+2}{x+1} = \frac{37}{6}$$

So we have

$$\frac{x+1}{x+2} = 6$$

$$x+1 = 6x+12$$

$$x-6x = 12-1$$

$$x = \frac{-11}{5}$$

Or

$$\frac{x+1}{x+2} = \frac{1}{6}$$

$$6x+6 = x+2$$

$$5x = -4$$

$$x = \frac{-4}{5}$$

So our answers are $x = \frac{-11}{5}$ or $x = \frac{-4}{5}$.

Now let's move on to our second case.

Case B

Let's start by taking an example, $x+\dfrac{1}{x} = \dfrac{13}{6}$.

Just keep in mind that the right-hand side is a little different here. Now to solve this, we'll take the factors of 6, which are 1, 2, 3 and 6.

Note that we cannot take 6 and $\dfrac{1}{6}$ here, because adding them would give $\dfrac{37}{6}$. So instead of 6 and 1, we will use 2 and 3 as reciprocals—that is, $\dfrac{2}{3} + \dfrac{3}{2}$ —then it would give us $\dfrac{13}{6}$.

So we now have

$$x + \frac{1}{x} = \frac{13}{6} = \frac{2}{3} + \frac{3}{2}$$

Here $x = \frac{2}{3}$ or $\frac{3}{2}$.

As I have been insisting, repetition is the secret to good memory. So let's repeat and understand the method better.

For our next example, let's take $x + \frac{1}{x} = \frac{25}{12}$.

First we'll try to find the factors of 12, so what we get is 1, 2, 3, 4, 6 and 12. Now if we take the factors 4 and 3 as reciprocals, $\frac{4}{3} + \frac{3}{4}$, then it would give us $\frac{25}{12}$. This means

$$x + \frac{1}{x} = \frac{25}{12} = \frac{4}{3} + \frac{3}{4}$$

Here $x = \frac{4}{3}$ or $\frac{3}{4}$, and we're done!

Let's take another example,

$$\frac{x+5}{x+6} + \frac{x+6}{x+5} = \frac{29}{10}$$

Here the right-hand-side denominator is 10. The factors are 1, 2, 5 and 10. We cannot use 1 and 10, because they don't give us $\frac{29}{10}$. So we use $\frac{2}{5}$ and $\frac{5}{2}$, which gives us $\frac{29}{10}$.

$$\frac{x+5}{x+6} + \frac{x+6}{x+5} = \frac{29}{10} = \frac{5}{2} + \frac{2}{5}$$

$$\frac{x+5}{x+6} = \frac{5}{2}$$

$$2x+10 = 5x+30$$

$$2x-5x = 30-10$$

$$x = \frac{-20}{3}$$

Or

$$\frac{x+5}{x+6} = \frac{2}{5}$$

$$5x+25 = 2x+12$$

$$5x-2x = 12-25$$

$$x = \frac{-13}{3}$$

So the values of x is $\frac{-20}{3}$ or $\frac{-13}{3}$. Wasn't that super quick?

Let's keep up the good work and solve another example:

Let's take $\frac{2x+11}{2x-11} + \frac{2x-11}{2x-11} = \frac{193}{84}$.

Here the right-hand-side denominator is 84. The factors are 1, 2, 3, 4, 6, 7, 12, 21, 28, 42 and 84. So we use $\frac{7}{12}$ and $\frac{12}{7}$, which gives us $\frac{193}{84}$.

Here the right-hand side is $\frac{7}{12} + \frac{12}{7}$.

So we'll solve it like this:

$$\frac{2x+11}{2x-11} = \frac{7}{12}$$

$$24x + 132 = 14x - 77$$

$$10x = -209$$

$$x = \frac{-209}{10}$$

Or

$$\frac{2x+11}{2x-11} = \frac{12}{7}$$

$$14x + 77 = 24x - 132$$

$$-10x = -132 - 77$$

$$x = \frac{209}{10}$$

That gives us the value of x as $\frac{209}{10}$ or $\frac{-209}{10}$.

Now let's look at the next case.

Case C

Let's take the example $x - \dfrac{1}{x} = \dfrac{5}{6}$.

Here again we have to be careful, because there is a minus sign. So we find the factors of 6, which are 1, 2, 3 and 6. We can't use 1 and 6, because they don't give us $\dfrac{5}{6}$; so we take 2 and 3 as reciprocals. Since $\dfrac{3}{2}$ is greater than $\dfrac{2}{3}$, we put that first and we have the right-hand side as $\dfrac{3}{2} - \dfrac{2}{3}$.

Note that the symbol is minus. So here we'll split the right-hand side like this:

$$x - \frac{1}{x} = \frac{5}{6} = \frac{3}{2} - \frac{2}{3}$$

The minus sign is important here. So we'll get $x = \frac{3}{2}$ or $\frac{-2}{3}$.

Let's take another similar example,

$$x - \frac{1}{x} = \frac{45}{14}$$

Here again we find the factors of 14, which are 1, 2, 7 and 14. We can't take 1 and 14, because they won't give us $\frac{45}{14}$, so instead, we go for 2 and 7 as reciprocals.

Since $\frac{7}{2}$ is greater than $\frac{2}{7}$, we put that first and we have the right-hand side as $\frac{7}{2} - \frac{2}{7}$.

So now our sum looks like this:

$$x - \frac{1}{x} = \frac{45}{14} = \frac{7}{2} - \frac{2}{7}$$

$$x = \frac{7}{2} \text{ or } \frac{-2}{7}$$

Now say we have

$$\frac{x}{x+3} - \frac{x+3}{x} = \frac{15}{56}$$

Here the factors of 56 are 1, 2, 4, 8, 7, 28 and 14. Out of

these, we take 7 and 8, because their reciprocals give us $\frac{15}{56}$.
So we have the right-hand side as $\frac{8}{7} - \frac{7}{8}$.

Following the same pattern as in the previous sums, we'll
solve like this:

$$\frac{x}{x+3} - \frac{x+3}{x} = \frac{15}{56} = \frac{8}{7} - \frac{7}{8}$$

And for x,

$$\frac{x}{x+3} = \frac{8}{7}$$

$$7x = 8x + 24$$

$$x = -24$$

Or

$$\frac{x}{x+3} = \frac{7}{8}$$

$$8x = 7x - 21$$

$$x = 21$$

So the values of x are -24 or 21.

Now let's take our final example,

$$\frac{5x+9}{5x-9} - \frac{5x-9}{5x+9} = \frac{56}{45}$$

Here the factors of 45 are 1, 3, 5, 9 and 15. We use 9 and
5, because their reciprocals subtracted give us $\frac{56}{45}$.

Here we'll split the right-hand side into two reciprocals. So what we get is

$$\frac{5x+9}{5x-9} - \frac{5x-9}{5x+9} = \frac{56}{45} = \frac{9}{5} - \frac{5}{9}$$

We'll now solve to find the two values of x:

$$\frac{5x+9}{5x-9} = \frac{9}{5}$$

$$25x+45 = 45x-81$$

$$-20x = -81 - 45$$

$$x = \frac{-126}{-20} = \frac{63}{10}$$

Or

$$\frac{5x+9}{5x-9} = \frac{-5}{9}$$

$$45x+81 = -25x+45$$

$$70x = 45-81$$

$$x = -\frac{36}{70} = \frac{-18}{35}$$

So the two values of x are $\frac{63}{10}$ or $\frac{-18}{35}$.

Calendars

THIS METHOD WAS shown to me by the late Shakuntala Devi, and I reproduce it here for your benefit. In this chapter, we will be learning how to remember calendars of over 600 years in split seconds. After this, we will be able to recall which day of the week it was in any given year after 15 October 1582, when the present Gregorian calendar was instituted. For example, if we have to find out which day of the week was 15 August 1947, most of us would be scratching our heads. But with this method, we will be able to find out easily. So let's get started.

Calendars

To be able to do this, we will have to remember four tables.

The first is the month table (see next page).

Here you see that for every month there is a single-digit figure. For example, for January, the corresponding figure is 0; for November, the corresponding figure is 3; for May, the corresponding figure is 1.

Month Table

Month	
Jan	0
Feb	3
Mar	3
Apr	6
May	1
Jun	4
Jul	6
Aug	2
Sep	5
Oct	0
Nov	3
Dec	5

This table has to be memorized. An easier way to do this would be to learn it in groups of four.

So you have 0 3 3 6 – 1 4 6 2 – 5 0 3 5 – if you memorize and repeat this in your mind four or five times, you will remember it!

Let's move on to the next table — the year table.

Month Table		Year Table	
Jan	0	1900	0
Feb	3	1904	5
Mar	3	1908	3
Apr	6	1912	1
May	1	1916	6
Jun	4	1920	4
Jul	6	1924	2
Aug	2		
Sep	5		
Oct	0		
Nov	3		
Dec	5		

There is something interesting about the year table. It starts in 1900 and goes on till 1924. Thereafter, the table repeats itself every four years, which means if 1924 had 2 as the corresponding figure, then 1928 will have the figure 0, 1932 will have the figure 5, 1936 will have the figure 3 and so on.

Every new century, say twentieth or twenty-first, starts from the beginning of the table. For example, if we were calculating for 1800, the figure would be 0, 1804 would be 5, 1808 would be 3 and so on.

We will now move on to the day table, which looks like this:

Month Table		Year Table		Day Table	
Jan	0	1900	0	Sun	0
Feb	3	1904	5	Mon	1
Mar	3	1908	3	Tue	2
Apr	6	1912	1	Wed	3
May	1	1916	6	Thu	4
Jun	4	1920	4	Fri	5
Jul	6	1924	2	Sat	6
Aug	2				
Sep	5				
Oct	0				
Nov	3				
Dec	5				

This day table is easy to remember. Sunday is 0, Monday is 1, Tuesday is 2 and so on.

Finally, we come to the century table:

Month Table		Year Table		Day Table		Century Table	
Jan	0	1900	0	Sun	0	21st	-1
Feb	3	1904	5	Mon	1	20th	0
Mar	3	1908	3	Tue	2	19th	2
Apr	6	1912	1	Wed	3	18th	4
May	1	1916	6	Thu	4	17th	6
Jun	4	1920	4	Fri	5		
Jul	6	1924	2	Sat	6		
Aug	2						
Sep	5						
Oct	0						
Nov	3						
Dec	5						

In the century table, we see that the twenty-first century has -1 as the corresponding figure, the twentieth century has the figure 0, the nineteenth century has the figure 2, the eighteenth century has the figure 4 and so on.

Now let's take a few examples and try to solve them.

1. Find out which day of the week was 9 November 1932.

We first find the date figure, which is 9.

We then find the month, which is November. We see the month table and see that November is 3. So our month figure is 3.

The year is 1932. So we go to the year table and we see that 1924 is 2.

Now the table repeats itself, so 1928 will be 0 and 1932 will be 5. So our year figure is 5.

We now come to the century figure.

1932 is the twentieth century, so we have the century figure as 0.

Now we find the sum total of the date figure, month figure, year figure and century figure.

From the total, we subtract the nearest multiple of 7 less than or equal to the number.

In this case, it is 14.

17 - 14 = 3, which, in the day table, corresponds to Wednesday!

So 9 November 1932 was a Wednesday.

9 November 1932

Date fig.:	9
Month fig.:	3
Year fig.:	5
Century fig.:	0
Leap year Adj.:	0

17 - 14 = 3
Wednesday!

We have to remember something important regarding the year figures.

If the year falls after 1900 and is a leap year, and the months are January or February, then subtract 1 from the total.

2. Let's find out which day of the week was 24 February 1936.

Our date figure is 24.

The figure corresponding to February is 3. So our month figure is 3.

The figure corresponding to 1936 is 3. So our year figure is 3.

The century figure is zero, as it is the twentieth century.

And because 1936 is a leap year and the month is February, we subtract 1 from the total. So our leap year adjustment is -1.

The total of this becomes 29, from which we subtract the nearest multiple of 7 less than equal to 29, which is 28. 29 - 28 = 1, which means that 24 February 1936 was a Monday.

24 February 1936

Date fig.:	24
Month fig.:	3
Year fig.:	3
Century fig.:	+ 0
Leap year Adj.:	- 1

$$29 - 28 = 1$$
Monday!

How about taking the case of a non-leap year?

3. What day of the week was 18 May 1943?

The date figure is 18.

The month figure is 1.

Please note how the year figure is calculated here (see table on the next page).

The figure corresponding 1924 is 2. Now since the table repeats, 1928 will be 0, 1932 will be 5, 1936 will be 3, 1940 will be 1 and 1944 will be 6.

1943 will be 3 years more than 1940. So we add 3 to 1 and the year figure becomes 4.

Month Table		Year Table		Day Table		Century Table	
• Jan	0	1900	0	Sun	0	21st	-1
• Feb	3	1904	5	Mon	1	20th	0
• Mar	3	1908	3	Tue	2	19th	2
• Apr	6	1912	1	Wed	3	18th	4
• May	1	1916	6	Thu	4	17th	6
• Jun	4	1920	4	Fri	5		
• Jul	6	1924	2	Sat	6		
• Aug	2						
• Sep	5						
• Oct	0						
• Nov	3						
• Dec	5						

The century figure will be 0, since the year is in the twentieth century.

18 May 1943

Date fig.:	18
Month fig.:	1
Year fig.:	4
Century fig.:	+ 0
Leap year Adj.:	---

23 - 21 = 2
Tuesday!

Now we can calculate almost any given date in the past 600 years.

4. Let's find out which day of the week was India's Independence Day, 15 August 1947.

15 August 1947

Date fig.:	15
Month fig.:	2
Year fig.:	9
Century fig.:	+ 0
Leap year Adj.:	---

26 - 21 = 5
Friday!

Our date figure here is 15.

The month figure corresponding to August is 2.

The year figure will be the figure corresponding to 1944, which is 6 plus 3. The year figure becomes 9.

The century figure will be 0.

Our total becomes 26, from which we subtract the multiple of 7 less than equal to 26. This is 21. 26 - 21 = 5.

In the date table, 5 corresponds to a Friday. So 15 August 1947 was a Friday!

I hope the method is clear now.

Let's see some more dates.

5. Which day of the week was 7 October 1875?

Our date figure is 7.

The month figure corresponding to October is 0.

The figure corresponding to 1872 is 6. Three more years will make it 1875. So our year figure would be 6 + 3 = 9.

1875 is in the nineteenth century. The figure corresponding to nineteenth century is 2.

7 October 1875

Date fig.:	7
Month fig.:	0
Year fig.:	9
Century fig.:	+ 2
Leap year Adj.:	---

---

18 - 14 = 4
Thursday!

6. Which day of the week was 2 October 1869?

Our date figure is 2.

The month figure corresponding to October is 0.

The year figure for 1868 is 1. We add 1 to 1 as 1869 is 1 year more than 1868. So our year figure is 2 as well.

Finally, the century figure is 2.

Adding these, we get 6, which means 2 October 1869 was a Saturday.

2 October 1869

Date fig.:	2
Month fig.:	0
Year fig.:	2
Century fig.:	+ 2
Leap year Adj.:	---

6
Saturday!

7. Which day of the year was 7 April 2007?

Our date figure will be 7.

April corresponds to 6. So our month figure is 6.

2004 corresponds to 5. To this we add 3, as 2007 is three years more than 2004. So our year figure becomes 5 + 3 = 8.

Since 2007 is in the twenty-first century, we will get the
century figure to be -1.

This totals up to be 20. From 20, we subtract 14, which
is the nearest multiple of 7 less than equal to 20.

This gives us 6, which corresponds to Saturday.

So 7 April 2007 was a Saturday!

7 April 2007

Date fig.:	7
Month fig.:	6
Year fig.:	8
Century fig.:	- 1
Leap year Adj.:	---
	20 - 14 = 6
	Saturday!

In this chapter, we learnt how to calculate the day of the
week for any given date in the last 600 years. If you learn to
memorize the tables, I am sure you will be a walking–talking
human calendar, just like Shakuntala Devi.

Practice Exercises

Find the product.

1. 91 × 95	2. 93 × 93	3. 95 × 94	4. 96 × 92	5. 91 × 98

6. 97 × 97	7. 91 × 97	8. 96 × 97	9. 98 × 99	10. 97 × 96

11. 96 × 90	12. 94 × 93	13. 94 × 96	14. 98 × 97	15. 91 × 96

16. 98 × 93	17. 93 × 91	18. 90 × 92	19. 93 × 96	20. 98 × 95

21. 95 × 95	22. 97 × 98	23. 96 × 98	24. 90 × 94	25. 99 × 93

26. 92 × 98	27. 97 × 94	28. 98 × 90	29. 96 × 95	30. 98 × 96

31. 95 × 96	32. 93 × 90	33. 90 × 95	34. 99 × 91	35. 91 × 94

36. 97 × 92	37. 93 × 98	38. 96 × 91	39. 92 × 91	40. 98 × 94

41. 90 × 98	42. 94 × 98	43. 92 × 90	44. 96 × 94	45. 92 × 95

46. 92 × 94	47. 99 × 99	48. 95 × 98	49. 94 × 92	50. 93 × 92

Find the product.

1. 991
    × 996
    ———

2. 994
    × 995
    ———

3. 996
    × 993
    ———

4. 990
    × 990
    ———

5. 993
    × 992
    ———

6. 997
    × 993
    ———

7. 993
    × 994
    ———

8. 994
    × 994
    ———

9. 992
    × 992
    ———

10. 992
    × 993
    ———

11. 997
    × 994
    ———

12. 992
    × 994
    ———

13. 993
    × 993
    ———

14. 999
    × 993
    ———

15. 998
    × 998
    ———

16. 996
    × 991
    ———

17. 993
    × 998
    ———

18. 991
    × 991
    ———

19. 998
    × 991
    ———

20. 996
    × 997
    ———

21. 996
    × 995
    ———

22. 995
    × 998
    ———

23. 998
    × 992
    ———

24. 993
    × 996
    ———

25. 998
    × 990
    ———

26. 994
    × 992
    ———

27. 995
    × 996
    ———

28. 991
    × 993
    ———

29. 994
    × 996
    ———

30. 995
    × 990
    ———

31. 991
    × 994
    ———

32. 993
    × 991
    ———

33. 994
    × 990
    ———

34. 997
    × 990
    ———

35. 997
    × 995
    ———

36. 998
    × 994
    ———

37. 995
    × 995
    ———

38. 994
    × 999
    ———

39. 994
    × 993
    ———

40. 992
    × 996
    ———

41. 991
    × 998
    ———

42. 993
    × 997
    ———

43. 999
    × 998
    ———

44. 992
    × 997
    ———

45. 996
    × 990
    ———

46. 992
    × 995
    ———

47. 995
    × 997
    ———

48. 998
    × 997
    ———

49. 994
    × 997
    ———

50. 994
    × 991
    ———

Find the product.

1.   17
  × 15
	2. 18
	  × 14
		3. 19
		  × 13
			4. 19
			  × 18
				5. 12
				  × 17

6.   12
  × 12
	7. 13
	  × 16
		8. 16
		  × 14
			9. 19
			  × 11
				10. 12
				  × 13

11.   15
  × 15
	12. 13
	  × 17
		13. 16
		  × 11
			14. 17
			  × 11
				15. 12
				  × 11

16.   17
  × 14
	17. 12
	  × 19
		18. 17
		  × 13
			19. 12
			  × 14
				20. 15
				  × 19

21.   15
  × 16
	22. 13
	  × 12
		23. 19
		  × 12
			24. 14
			  × 19
				25. 18
				  × 15

26.   11
  × 12
	27. 16
	  × 18
		28. 16
		  × 12
			29. 14
			  × 18
				30. 16
				  × 17

31.   19
  × 15
	32. 18
	  × 12
		33. 17
		  × 18
			34. 12
			  × 16
				35. 18
				  × 19

36.   11
  × 14
	37. 13
	  × 18
		38. 17
		  × 16
			39. 11
			  × 13
				40. 13
				  × 14

41.   16
  × 15
	42. 11
	  × 15
		43. 14
		  × 17
			44. 16
			  × 13
				45. 14
				  × 12

46.   14
  × 13
	47. 18
	  × 17
		48. 19
		  × 16
			49. 18
			  × 11
				50. 12
				  × 18

Find the product.

1.   103
   × 102
	2. 112
	   × 113
		3. 116
		   × 110
			4. 120
			   × 113
				5. 121
				   × 108

6.   111
   × 102
	7. 115
	   × 115
		8. 102
		   × 118
			9. 105
			   × 100
				10. 102
				   × 109

11.   120
   × 116
	12. 107
	   × 112
		13. 120
		   × 117
			14. 117
			   × 119
				15. 113
				   × 105

16.   109
   × 115
	17. 115
	   × 109
		18. 116
		   × 114
			19. 115
			   × 104
				20. 114
				   × 108

21.   104
   × 120
	22. 105
	   × 111
		23. 118
		   × 114
			24. 113
			   × 121
				25. 101
				   × 110

26.   102
   × 106
	27. 107
	   × 108
		28. 117
		   × 121
			29. 114
			   × 120
				30. 114
				   × 101

31.   100
   × 116
	32. 108
	   × 106
		33. 120
		   × 109
			34. 103
			   × 100
				35. 109
				   × 110

36.   115
   × 111
	37. 112
	   × 108
		38. 107
		   × 109
			39. 120
			   × 103
				40. 107
				   × 114

41.   104
   × 119
	42. 113
	   × 109
		43. 102
		   × 108
			44. 118
			   × 115
				45. 112
				   × 112

46.   108
   × 109
	47. 121
	   × 109
		48. 103
		   × 111
			49. 115
			   × 103
				50. 118
				   × 101

Find the product.

1.   18
  × 8
	2. 13
	  × 7
		3. 19
		  × 6
			4. 12
			  × 9
				5. 11
				  × 8

6.   17
  × 8
	7. 18
	  × 9
		8. 16
		  × 8
			9. 17
			  × 9
				10. 16
				  × 9

11.   14
  × 6
	12. 15
	  × 9
		13. 11
		  × 6
			14. 13
			  × 8
				15. 14
				  × 8

16.   18
  × 6
	17. 14
	  × 7
		18. 16
		  × 7
			19. 13
			  × 6
				20. 17
				  × 6

21.   12
  × 7
	22. 15
	  × 6
		23. 15
		  × 8
			24. 12
			  × 6
				25. 12
				  × 8

26.   15
  × 7
	27. 16
	  × 6
		28. 11
		  × 7
			29. 19
			  × 7
				30. 18
				  × 7

31.   17
  × 7
	32. 13
	  × 9
		33. 14
		  × 9
			34. 19
			  × 9
				35. 19
				  × 8

36.   11
  × 9
	37. 12
	  × 8
		38. 12
		  × 8
			39. 15
			  × 9
				40. 14
				  × 7

41.   15
  × 6
	42. 15
	  × 8
		43. 17
		  × 7
			44. 12
			  × 6
				45. 13
				  × 6

46.   19
  × 7
	47. 14
	  × 8
		48. 17
		  × 6
			49. 18
			  × 8
				50. 15
				  × 7

Find the product.

1. 110 × 92	2. 116 × 90	3. 117 × 98	4. 101 × 95	5. 109 × 97
6. 113 × 91	7. 112 × 98	8. 107 × 92	9. 114 × 96	10. 114 × 92
11. 114 × 94	12. 112 × 91	13. 102 × 92	14. 118 × 92	15. 112 × 90
16. 110 × 94	17. 107 × 96	18. 111 × 98	19. 116 × 96	20. 118 × 91
21. 106 × 97	22. 107 × 97	23. 109 × 91	24. 111 × 90	25. 110 × 97
26. 106 × 93	27. 113 × 96	28. 118 × 96	29. 118 × 97	30. 118 × 94
31. 106 × 92	32. 108 × 94	33. 108 × 90	34. 115 × 95	35. 118 × 98
36. 104 × 97	37. 114 × 98	38. 113 × 98	39. 103 × 96	40. 105 × 91
41. 115 × 92	42. 104 × 94	43. 111 × 92	44. 107 × 95	45. 113 × 90
46. 111 × 96	47. 115 × 97	48. 101 × 90	49. 119 × 91	50. 102 × 91

Find the product.

1. 87
    × 27
    ───

2. 21
    × 35
    ───

3. 14
    × 31
    ───

4. 30
    × 40
    ───

5. 42
    × 32
    ───

6. 88
    × 58
    ───

7. 65
    × 18
    ───

8. 78
    × 45
    ───

9. 13
    × 60
    ───

10. 46
    × 76
    ───

11. 74
    × 97
    ───

12. 56
    × 65
    ───

13. 52
    × 68
    ───

14. 39
    × 65
    ───

15. 31
    × 33
    ───

16. 79
    × 73
    ───

17. 19
    × 46
    ───

18. 48
    × 71
    ───

19. 47
    × 29
    ───

20. 38
    × 13
    ───

21. 77
    × 24
    ───

22. 81
    × 61
    ───

23. 85
    × 57
    ───

24. 41
    × 31
    ───

25. 56
    × 44
    ───

26. 66
    × 23
    ───

27. 84
    × 30
    ───

28. 50
    × 89
    ───

29. 70
    × 56
    ───

30. 82
    × 46
    ───

31. 91
    × 65
    ───

32. 20
    × 20
    ───

33. 51
    × 25
    ───

34. 50
    × 30
    ───

35. 68
    × 19
    ───

36. 97
    × 83
    ───

37. 34
    × 44
    ───

38. 98
    × 61
    ───

39. 27
    × 94
    ───

40. 79
    × 94
    ───

41. 38
    × 18
    ───

42. 20
    × 86
    ───

43. 66
    × 67
    ───

44. 88
    × 87
    ───

45. 95
    × 58
    ───

46. 12
    × 90
    ───

47. 46
    × 91
    ───

48. 21
    × 23
    ───

49. 15
    × 71
    ───

50. 20
    × 71
    ───

Find the product.

1. 894 × 114	2. 460 × 358	3. 981 × 801	4. 946 × 548	5. 608 × 483
6. 164 × 745	7. 816 × 463	8. 252 × 156	9. 375 × 914	10. 941 × 577
11. 655 × 369	12. 767 × 398	13. 966 × 846	14. 607 × 671	15. 702 × 160
16. 893 × 814	17. 943 × 816	18. 239 × 996	19. 774 × 815	20. 949 × 240
21. 766 × 801	22. 757 × 142	23. 278 × 873	24. 629 × 320	25. 620 × 366
26. 585 × 435	27. 206 × 426	28. 778 × 626	29. 346 × 163	30. 391 × 237
31. 295 × 376	32. 845 × 549	33. 590 × 349	34. 802 × 102	35. 326 × 255
36. 381 × 559	37. 963 × 802	38. 810 × 853	39. 611 × 726	40. 221 × 193
41. 317 × 388	42. 151 × 437	43. 574 × 559	44. 709 × 617	45. 139 × 612
46. 750 × 712	47. 525 × 814	48. 944 × 491	49. 955 × 309	50. 551 × 929

Find the sum.

1.	76 + 37	2.	47 + 19	3.	53 + 25	4.	13 + 61	5.	62 + 30
6.	51 + 45	7.	97 + 15	8.	34 + 34	9.	85 + 89	10.	29 + 49
11.	62 + 41	12.	23 + 25	13.	39 + 84	14.	41 + 83	15.	23 + 71
16.	64 + 48	17.	82 + 45	18.	18 + 86	19.	80 + 30	20.	44 + 16
21.	17 + 45	22.	35 + 70	23.	15 + 75	24.	16 + 36	25.	41 + 40
26.	58 + 26	27.	15 + 67	28.	96 + 45	29.	24 + 32	30.	79 + 68
31.	83 + 29	32.	48 + 90	33.	64 + 13	34.	62 + 62	35.	96 + 47
36.	54 + 65	37.	55 + 23	38.	42 + 27	39.	16 + 52	40.	20 + 63
41.	78 + 15	42.	52 + 82	43.	23 + 48	44.	43 + 25	45.	48 + 72
46.	72 + 58	47.	50 + 75	48.	32 + 87	49.	95 + 44	50.	16 + 12

Find the sum.

1.	193 + 149	2.	439 + 907	3.	711 + 996	4.	933 + 409	5.	897 + 834
6.	212 + 122	7.	446 + 371	8.	588 + 584	9.	972 + 285	10.	891 + 449
11.	871 + 474	12.	794 + 577	13.	995 + 922	14.	921 + 380	15.	233 + 591
16.	286 + 374	17.	158 + 641	18.	699 + 726	19.	593 + 620	20.	132 + 453
21.	313 + 684	22.	943 + 240	23.	421 + 805	24.	607 + 401	25.	318 + 447
26.	609 + 194	27.	521 + 488	28.	285 + 719	29.	354 + 862	30.	910 + 787
31.	175 + 654	32.	278 + 407	33.	649 + 837	34.	963 + 804	35.	671 + 861
36.	990 + 453	37.	254 + 539	38.	236 + 287	39.	529 + 655	40.	256 + 291
41.	416 + 622	42.	163 + 785	43.	981 + 193	44.	852 + 669	45.	457 + 979
46.	474 + 731	47.	709 + 344	48.	379 + 659	49.	508 + 637	50.	590 + 583

Find the sum.

1. 6,417 + 8,351	2. 9,823 + 2,251	3. 2,626 + 6,179	4. 7,793 + 2,657	5. 8,003 + 6,008
6. 1,138 + 1,542	7. 5,860 + 1,802	8. 3,615 + 6,273	9. 5,985 + 2,208	10. 5,158 + 3,666
11. 8,182 + 2,668	12. 9,697 + 9,632	13. 9,145 + 8,952	14. 1,291 + 7,363	15. 4,876 + 4,988
16. 3,668 + 4,924	17. 7,335 + 3,441	18. 3,264 + 3,058	19. 8,419 + 3,735	20. 1,474 + 6,912
21. 7,838 + 2,179	22. 8,060 + 6,202	23. 7,153 + 2,913	24. 4,147 + 5,120	25. 9,259 + 2,912
26. 4,221 + 2,672	27. 4,037 + 3,277	28. 4,050 + 4,759	29. 8,653 + 1,201	30. 3,105 + 3,673
31. 5,898 + 1,441	32. 7,780 + 4,025	33. 3,350 + 6,213	34. 4,205 + 3,428	35. 8,537 + 1,568
36. 7,854 + 3,261	37. 1,744 + 8,245	38. 5,827 + 7,012	39. 4,608 + 4,505	40. 3,489 + 7,242
41. 5,815 + 6,170	42. 8,102 + 2,974	43. 1,777 + 2,006	44. 1,898 + 5,068	45. 9,390 + 8,504
46. 4,404 + 8,228	47. 1,640 + 7,800	48. 2,024 + 7,407	49. 4,398 + 5,045	50. 3,862 + 1,951

Gaurav Tekriwal

Find the difference.

1.   75
   - 60
	2. 78
	   - 78
		3. 56
		   - 42
			4. 89
			   - 21
				5. 78
				   - 39

6.   90
   - 56
	7. 65
	   - 65
		8. 88
		   - 45
			9. 61
			   - 53
				10. 85
				   - 47

11. 33
   - 15
	12. 25
	   - 13
		13. 77
		   - 49
			14. 52
			   - 22
				15. 34
				   - 22

16. 60
   - 16
	17. 63
	   - 52
		18. 24
		   - 24
			19. 24
			   - 16
				20. 94
				   - 18

21. 54
   - 35
	22. 63
	   - 62
		23. 75
		   - 39
			24. 82
			   - 28
				25. 64
				   - 31

26. 74
   - 15
	27. 18
	   - 13
		28. 92
		   - 20
			29. 51
			   - 29
				30. 69
				   - 18

31. 54
   - 48
	32. 94
	   - 46
		33. 75
		   - 72
			34. 88
			   - 63
				35. 89
				   - 85

36. 98
   - 40
	37. 83
	   - 57
		38. 90
		   - 25
			39. 95
			   - 89
				40. 63
				   - 63

41. 84
   - 58
	42. 40
	   - 34
		43. 91
		   - 69
			44. 85
			   - 83
				45. 94
				   - 78

46. 56
   - 17
	47. 64
	   - 33
		48. 56
		   - 27
			49. 70
			   - 66
				50. 69
				   - 17

Find the difference.

1. 974
   - 820
	2. 702
	   - 448
		3. 460
		   - 283
			4. 596
			   - 435
				5. 834
				   - 258

6. 964
   - 920
	7. 950
	   - 649
		8. 634
		   - 606
			9. 737
			   - 356
				10. 751
				    - 369

11. 623
    - 583
	12. 527
	    - 333
		13. 916
		    - 297
			14. 414
			    - 233
				15. 973
				    - 407

16. 514
    - 122
	17. 989
	    - 632
		18. 860
		    - 657
			19. 720
			    - 296
				20. 755
				    - 363

21. 843
    - 769
	22. 751
	    - 460
		23. 840
		    - 835
			24. 983
			    - 219
				25. 835
				    - 271

26. 336
    - 251
	27. 604
	    - 451
		28. 788
		    - 566
			29. 316
			    - 303
				30. 905
				    - 136

31. 873
    - 437
	32. 961
	    - 884
		33. 312
		    - 116
			34. 527
			    - 347
				35. 970
				    - 778

36. 644
    - 460
	37. 962
	    - 799
		38. 748
		    - 228
			39. 760
			    - 439
				40. 691
				    - 642

41. 981
    - 450
	42. 723
	    - 357
		43. 846
		    - 464
			44. 333
			    - 227
				45. 927
				    - 372

46. 922
    - 728
	47. 825
	    - 107
		48. 641
		    - 269
			49. 640
			    - 102
				50. 478
				    - 160

Gaurav Tekriwal

Find the difference.

1.   3,777
  - 1,569
	2. 7,524
	  - 5,559
		3. 6,772
		  - 6,676
			4. 9,710
			  - 6,211
				5. 6,867
				  - 6,021

6.   7,236
  - 5,604
	7. 5,988
	  - 4,181
		8. 9,714
		  - 4,272
			9. 4,348
			  - 2,034
				10. 6,177
				  - 1,700

11.   5,706
  - 3,587
	12. 5,353
	  - 5,338
		13. 9,560
		  - 1,685
			14. 9,913
			  - 2,261
				15. 5,580
				  - 3,855

16.   7,111
  - 3,278
	17. 4,390
	  - 2,662
		18. 8,973
		  - 8,597
			19. 2,291
			  - 1,642
				20. 5,909
				  - 1,474

21.   9,897
  - 3,401
	22. 4,711
	  - 3,062
		23. 5,948
		  - 5,325
			24. 9,034
			  - 1,493
				25. 6,727
				  - 4,711

26.   5,561
  - 2,237
	27. 4,268
	  - 3,946
		28. 8,274
		  - 3,180
			29. 6,152
			  - 2,736
				30. 8,566
				  - 2,328

31.   6,713
  - 4,648
	32. 8,930
	  - 1,803
		33. 8,457
		  - 5,630
			34. 8,100
			  - 6,853
				35. 8,786
				  - 7,120

36.   5,191
  - 2,642
	37. 7,500
	  - 4,710
		38. 9,266
		  - 2,045
			39. 6,519
			  - 2,328
				40. 9,344
				  - 8,605

41.   5,846
  - 2,422
	42. 4,393
	  - 1,382
		43. 7,650
		  - 7,251
			44. 7,544
			  - 6,924
				45. 4,826
				  - 3,003

46.   2,777
  - 1,166
	47. 5,727
	  - 2,603
		48. 8,808
		  - 7,054
			49. 9,521
			  - 8,821
				50. 8,031
				  - 7,570

Find the quotient.

1. 31 ⟌ 9,517

2. 86 ⟌ 7,052

3. 47 ⟌ 1,927

4. 20 ⟌ 1,700

5. 34 ⟌ 1,224

6. 28 ⟌ 9,800

7. 44 ⟌ 7,392

8. 49 ⟌ 4,606

9. 42 ⟌ 9,912

10. 68 ⟌ 3,944

11. 42 ⟌ 714

12. 45 ⟌ 675

13. 67 ⟌ 8,643

14. 68 ⟌ 9,928

15. 30 ⟌ 6,420

16. 93 ⟌ 6,789

17. 22 ⟌ 6,776

18. 74 ⟌ 9,546

19. 24 ⟌ 240

20. 66 ⟌ 1,122

21. 54 ⟌ 7,722

22. 48 ⟌ 1,200

23. 38 ⟌ 4,750

24. 80 ⟌ 320

25. 13 ⟌ 2,509

26. 46 ⟌ 9,844

27. 44 ⟌ 6,336

28. 63 ⟌ 1,512

29. 38 ⟌ 1,786

30. 59 ⟌ 9,912

31. 55 ⟌ 4,235

32. 89 ⟌ 4,717

33. 14 ⟌ 5,894

34. 64 ⟌ 5,248

35. 75 ⟌ 4,275

36. 85 ⟌ 4,335

37. 78 ⟌ 2,652

38. 61 ⟌ 8,784

39. 66 ⟌ 2,508

40. 59 ⟌ 9,204

41. 29 ⟌ 6,235

42. 64 ⟌ 4,096

43. 34 ⟌ 2,414

44. 36 ⟌ 3,024

45. 42 ⟌ 9,786

46. 73 ⟌ 5,548

47. 60 ⟌ 1,680

48. 34 ⟌ 2,142

49. 30 ⟌ 240

50. 47 ⟌ 8,460

Find the sum.

1. $\frac{3}{4} + \frac{1}{14} =$ ___ 2. $\frac{1}{7} + \frac{9}{17} =$ ___ 3. $\frac{12}{13} + \frac{8}{15} =$ ___ 4. $\frac{4}{5} + \frac{14}{18} =$ ___ 5. $\frac{21}{25} + \frac{3}{4} =$ ___

6. $\frac{2}{12} + \frac{6}{9} =$ ___ 7. $\frac{10}{11} + \frac{5}{20} =$ ___ 8. $\frac{3}{10} + \frac{2}{6} =$ ___ 9. $\frac{6}{20} + \frac{1}{10} =$ ___ 10. $\frac{2}{3} + \frac{4}{9} =$ ___

11. $\frac{5}{14} + \frac{6}{20} =$ ___ 12. $\frac{8}{17} + \frac{10}{11} =$ ___ 13. $\frac{3}{8} + \frac{16}{17} =$ ___ 14. $\frac{1}{9} + \frac{2}{16} =$ ___ 15. $\frac{1}{3} + \frac{3}{5} =$ ___

16. $\frac{15}{16} + \frac{10}{14} =$ ___ 17. $\frac{1}{15} + \frac{9}{10} =$ ___ 18. $\frac{6}{12} + \frac{1}{4} =$ ___ 19. $\frac{5}{30} + \frac{10}{16} =$ ___ 20. $\frac{2}{3} + \frac{4}{5} =$ ___

21. $\frac{13}{17} + \frac{6}{8} =$ ___ 22. $\frac{5}{12} + \frac{2}{3} =$ ___ 23. $\frac{5}{20} + \frac{5}{15} =$ ___ 24. $\frac{16}{18} + \frac{2}{4} =$ ___ 25. $\frac{1}{8} + \frac{14}{18} =$ ___

26. $\frac{24}{25} + \frac{2}{5} =$ ___ 27. $\frac{8}{16} + \frac{1}{3} =$ ___ 28. $\frac{7}{14} + \frac{1}{16} =$ ___ 29. $\frac{28}{40} + \frac{12}{14} =$ ___ 30. $\frac{2}{8} + \frac{10}{11} =$ ___

31. $\frac{5}{7} + \frac{13}{17} =$ ___ 32. $\frac{2}{3} + \frac{5}{11} =$ ___ 33. $\frac{3}{4} + \frac{3}{17} =$ ___ 34. $\frac{5}{30} + \frac{7}{18} =$ ___ 35. $\frac{10}{14} + \frac{2}{9} =$ ___

36. $\frac{21}{25} + \frac{2}{7} =$ ___ 37. $\frac{4}{5} + \frac{3}{5} =$ ___ 38. $\frac{9}{17} + \frac{14}{20} =$ ___ 39. $\frac{8}{15} + \frac{3}{4} =$ ___ 40. $\frac{15}{16} + \frac{4}{6} =$ ___

41. $\frac{1}{6} + \frac{9}{16} =$ ___ 42. $\frac{2}{3} + \frac{9}{17} =$ ___ 43. $\frac{4}{14} + \frac{8}{12} =$ ___ 44. $\frac{4}{11} + \frac{5}{11} =$ ___ 45. $\frac{3}{7} + \frac{2}{9} =$ ___

46. $\frac{18}{40} + \frac{14}{18} =$ ___ 47. $\frac{1}{8} + \frac{1}{18} =$ ___ 48. $\frac{5}{17} + \frac{1}{7} =$ ___ 49. $\frac{1}{4} + \frac{1}{5} =$ ___ 50. $\frac{3}{15} + \frac{6}{11} =$ ___

Find the difference.

1. $\frac{23}{30} - \frac{3}{14} =$ ___   2. $\frac{5}{7} - \frac{6}{50} =$ ___   3. $\frac{14}{18} - \frac{1}{3} =$ ___   4. $\frac{6}{10} - \frac{4}{14} =$ ___   5. $\frac{27}{50} - \frac{15}{50} =$ ___

6. $\frac{8}{11} - \frac{3}{6} =$ ___   7. $\frac{13}{18} - \frac{1}{4} =$ ___   8. $\frac{6}{9} - \frac{2}{40} =$ ___   9. $\frac{3}{4} - \frac{7}{10} =$ ___   10. $\frac{6}{7} - \frac{5}{18} =$ ___

11. $\frac{3}{6} - \frac{14}{30} =$ ___   12. $\frac{10}{14} - \frac{13}{50} =$ ___   13. $\frac{5}{9} - \frac{3}{14} =$ ___   14. $\frac{11}{12} - \frac{4}{17} =$ ___   15. $\frac{11}{20} - \frac{4}{12} =$ ___

16. $\frac{12}{25} - \frac{11}{25} =$ ___   17. $\frac{5}{18} - \frac{13}{50} =$ ___   18. $\frac{10}{12} - \frac{5}{8} =$ ___   19. $\frac{5}{9} - \frac{16}{40} =$ ___   20. $\frac{12}{14} - \frac{7}{14} =$ ___

21. $\frac{14}{16} - \frac{19}{30} =$ ___   22. $\frac{5}{40} - \frac{2}{25} =$ ___   23. $\frac{8}{9} - \frac{4}{5} =$ ___   24. $\frac{12}{25} - \frac{1}{4} =$ ___   25. $\frac{15}{20} - \frac{1}{8} =$ ___

26. $\frac{1}{2} - \frac{22}{50} =$ ___   27. $\frac{6}{25} - \frac{2}{12} =$ ___   28. $\frac{3}{11} - \frac{1}{14} =$ ___   29. $\frac{5}{18} - \frac{1}{20} =$ ___   30. $\frac{6}{7} - \frac{15}{18} =$ ___

31. $\frac{4}{15} - \frac{1}{9} =$ ___   32. $\frac{17}{18} - \frac{18}{25} =$ ___   33. $\frac{5}{6} - \frac{14}{17} =$ ___   34. $\frac{28}{40} - \frac{2}{40} =$ ___   35. $\frac{4}{6} - \frac{2}{6} =$ ___

36. $\frac{3}{11} - \frac{3}{16} =$ ___   37. $\frac{5}{9} - \frac{4}{18} =$ ___   38. $\frac{2}{3} - \frac{4}{9} =$ ___   39. $\frac{4}{11} - \frac{1}{5} =$ ___   40. $\frac{6}{7} - \frac{8}{12} =$ ___

41. $\frac{23}{40} - \frac{3}{6} =$ ___   42. $\frac{3}{12} - \frac{1}{30} =$ ___   43. $\frac{10}{16} - \frac{5}{50} =$ ___   44. $\frac{16}{18} - \frac{3}{10} =$ ___   45. $\frac{6}{11} - \frac{2}{10} =$ ___

46. $\frac{10}{11} - \frac{7}{20} =$ ___   47. $\frac{4}{5} - \frac{13}{25} =$ ___   48. $\frac{14}{15} - \frac{4}{6} =$ ___   49. $\frac{11}{16} - \frac{20}{30} =$ ___   50. $\frac{8}{9} - \frac{1}{18} =$ ___

Find the sum.

1. 3.832 + 7.909	2. 927.3 + 556.5	3. 597.4 + 269.7	4. 4.082 + 9.356	5. 156.0 + 485.5
6. 158.4 + 847.9	7. 9.164 + 8.544	8. 3,667 + 7,780	9. 41.51 + 70.44	10. 92.07 + 70.86
11. 2,081 + 1,039	12. 95.82 + 87.90	13. 272.7 + 236.1	14. 48.01 + 54.28	15. 8.601 + 6.936
16. 9.972 + 4.593	17. 490.3 + 334.2	18. 8.619 + 6.017	19. 27.76 + 88.40	20. 2.519 + 3.751
21. 79.58 + 57.59	22. 22.42 + 49.01	23. 7,435 + 7,219	24. 18.65 + 53.53	25. 76.62 + 97.99
26. 63.25 + 31.26	27. 33.47 + 31.42	28. 27.81 + 37.94	29. 832.5 + 148.4	30. 38.39 + 46.14
31. 78.94 + 86.53	32. 17.57 + 46.24	33. 48.82 + 69.89	34. 4,577 + 6,823	35. 9,924 + 7,649
36. 6.004 + 5.310	37. 5,526 + 6,457	38. 91.27 + 63.21	39. 879.6 + 568.3	40. 2,356 + 5,507
41. 1.497 + 9.447	42. 776.6 + 880.1	43. 1.611 + 7.846	44. 224.7 + 276.1	45. 15.62 + 58.26
46. 769.1 + 229.3	47. 6.002 + 7.905	48. 5.855 + 9.304	49. 50.04 + 76.17	50. 1.019 + 1.446

Find the difference.

1. 930.9
  - 372.9
	2. 939.8
	  - 795.1
		3. 7,753
		  - 1,031
			4. 9,832
			  - 6,398
				5. 6,798
				  - 2,159

6. 7,235
  - 4,055
	7. 8,090
	  - 4,490
		8. 96.77
		  - 46.83
			9. 8,553
			  - 8,341
				10. 53.16
				  - 42.69

11. 623.8
  - 515.2
	12. 9,498
	  - 5,768
		13. 7,676
		  - 4,446
			14. 59.98
			  - 59.32
				15. 899.2
				  - 474.1

16. 62.47
  - 58.35
	17. 479.3
	  - 350.7
		18. 86.65
		  - 83.16
			19. 59.20
			  - 17.76
				20. 8,056
				  - 3,159

21. 5,690
  - 4,812
	22. 7,249
	  - 7,132
		23. 94.74
		  - 37.32
			24. 8,183
			  - 5,547
				25. 75.53
				  - 47.82

26. 6,893
  - 6,345
	27. 752.7
	  - 585.0
		28. 73.18
		  - 31.37
			29. 405.5
			  - 318.9
				30. 922.6
				  - 363.5

31. 8,009
  - 1,376
	32. 830.9
	  - 220.4
		33. 88.80
		  - 74.73
			34. 79.18
			  - 67.49
				35. 9,001
				  - 2,400

36. 35.79
  - 12.80
	37. 181.0
	  - 132.3
		38. 4,756
		  - 1,784
			39. 6,824
			  - 5,750
				40. 9,969
				  - 8,972

41. 967.1
  - 512.7
	42. 9,954
	  - 3,535
		43. 83.49
		  - 35.53
			44. 766.7
			  - 157.0
				45. 2,065
				  - 1,096

46. 70.86
  - 28.00
	47. 29.53
	  - 25.41
		48. 5,379
		  - 4,131
			49. 63.51
			  - 17.42
				50. 996.3
				  - 827.3

Find the product.

1.	59 × 31	2.	0.29 × 7.9	3.	86 × 0.54	4.	4.7 × 1.6	5.	9.7 × 0.28
6.	0.64 × 8.0	7.	61 × 12	8.	0.77 × 6.5	9.	8.8 × 0.52	10.	0.38 × 5.1
11.	5.8 × 29	12.	2.0 × 0.43	13.	0.78 × 0.78	14.	31 × 3.6	15.	5.3 × 0.68
16.	28 × 39	17.	2.8 × 9.3	18.	6.8 × 0.56	19.	8.5 × 5.6	20.	15 × 32
21.	0.64 × 7.6	22.	49 × 82	23.	63 × 78	24.	53 × 56	25.	8.4 × 5.9
26.	0.30 × 38	27.	5.2 × 44	28.	0.93 × 0.31	29.	0.67 × 0.67	30.	0.57 × 30
31.	0.67 × 94	32.	0.14 × 0.44	33.	0.96 × 0.66	34.	55 × 2.9	35.	7.7 × 3.7
36.	1.2 × 54	37.	95 × 0.81	38.	6.0 × 7.5	39.	4.0 × 0.13	40.	8.9 × 91
41.	0.58 × 0.81	42.	0.29 × 0.93	43.	0.12 × 0.45	44.	0.33 × 7.9	45.	0.10 × 5.6
46.	0.36 × 17	47.	5.6 × 26	48.	0.59 × 0.89	49.	88 × 0.98	50.	3.3 × 5.9

Find the product.

1. 5.47
    × 43.1
    ———

2. 675
    × 39.6
    ———

3. 819
    × 226
    ———

4. 45.6
    × 53.5
    ———

5. 3.83
    × 10.6
    ———

6. 9.86
    × 6.00
    ———

7. 7.93
    × 5.08
    ———

8. 9.11
    × 2.15
    ———

9. 132
    × 175
    ———

10. 24.5
    × 700
    ———

11. 208
    × 58.1
    ———

12. 78.2
    × 104
    ———

13. 9.81
    × 6.87
    ———

14. 74.1
    × 4.08
    ———

15. 331
    × 80.6
    ———

16. 20.0
    × 27.5
    ———

17. 611
    × 286
    ———

18. 9.32
    × 3.57
    ———

19. 153
    × 63.6
    ———

20. 97.0
    × 41.9
    ———

21. 660
    × 680
    ———

22. 765
    × 37.3
    ———

23. 200
    × 825
    ———

24. 8.09
    × 31.5
    ———

25. 6.45
    × 82.6
    ———

26. 53.6
    × 6.63
    ———

27. 2.50
    × 437
    ———

28. 63.8
    × 965
    ———

29. 64.8
    × 6.72
    ———

30. 414
    × 23.3
    ———

31. 48.4
    × 1.48
    ———

32. 30.5
    × 576
    ———

33. 90.5
    × 812
    ———

34. 98.8
    × 6.08
    ———

35. 5.76
    × 655
    ———

36. 163
    × 4.95
    ———

37. 2.03
    × 82.5
    ———

38. 14.1
    × 527
    ———

39. 81.5
    × 510
    ———

40. 903
    × 25.6
    ———

41. 8.77
    × 527
    ———

42. 76.2
    × 50.4
    ———

43. 3.71
    × 536
    ———

44. 126
    × 27.5
    ———

45. 647
    × 96.6
    ———

46. 708
    × 4.80
    ———

47. 323
    × 83.8
    ———

48. 9.34
    × 61.2
    ———

49. 47.6
    × 2.90
    ———

50. 467
    × 38.7
    ———

Find the quotient.

1. 5.5) 9.89

2. 0.17) 90.8

3. 0.39) 23.0

4. 0.62) 36.2

5. 1.5) 1.64

6. 0.98) 60.0

7. 0.66) 53.4

8. 9.3) 79.0

9. 1.6) 2.91

10. 6.0) 61.3

11. 0.35) 77.4

12. 2.9) 2.23

13. 0.98) 2.96

14. 0.48) 7.21

15. 8.2) 2.50

16. 4.8) 2.12

17. 0.91) 23.4

18. 4.5) 65.1

19. 0.38) 4.72

20. 6.7) 82.7

21. 7.9) 46.9

22. 0.36) 9.57

23. 1.1) 6.05

24. 0.78) 2.40

25. 4.6) 12.1

26. 9.2) 33.3

27. 0.63) 35.0

28. 7.3) 9.83

29. 7.1) 43.3

30. 0.23) 87.4

31. 8.7) 67.4

32. 0.33) 47.5

33. 2.2) 93.4

34. 0.61) 8.54

35. 4.4) 11.8

36. 0.90) 56.6

37. 0.93) 32.8

38. 9.7) 2.37

39. 8.5) 3.94

40. 2.8) 68.2

41. 9.3) 8.53

42. 0.78) 4.58

43. 0.28) 59.4

44. 0.59) 2.33

45. 9.0) 9.38

46. 0.74) 1.16

47. 0.69) 50.5

48. 0.58) 10.2

49. 7.1) 4.19

50. 4.8) 4.31

Convert to decimals.

1. $\frac{8}{19}$ = _____   2. $\frac{1}{7}$ = _____   3. $\frac{3}{17}$ = _____   4. $\frac{3}{21}$ = _____   5. $\frac{18}{23}$ = _____

6. $\frac{1}{3}$ = _____   7. $\frac{3}{13}$ = _____   8. $\frac{13}{19}$ = _____   9. $\frac{5}{11}$ = _____   10. $\frac{12}{13}$ = _____

11. $\frac{5}{21}$ = _____   12. $\frac{2}{19}$ = _____   13. $\frac{16}{17}$ = _____   14. $\frac{5}{7}$ = _____   15. $\frac{13}{23}$ = _____

16. $\frac{2}{3}$ = _____   17. $\frac{9}{13}$ = _____   18. $\frac{7}{11}$ = _____   19. $\frac{6}{17}$ = _____   20. $\frac{3}{7}$ = _____

21. $\frac{11}{21}$ = _____   22. $\frac{7}{23}$ = _____   23. $\frac{4}{19}$ = _____   24. $\frac{16}{21}$ = _____   25. $\frac{7}{19}$ = _____

26. $\frac{10}{11}$ = _____   27. $\frac{8}{23}$ = _____   28. $\frac{2}{13}$ = _____   29. $\frac{2}{11}$ = _____   30. $\frac{6}{7}$ = _____

31. $\frac{11}{19}$ = _____   32. $\frac{2}{17}$ = _____   33. $\frac{5}{13}$ = _____   34. $\frac{1}{11}$ = _____   35. $\frac{10}{19}$ = _____

36. $\frac{13}{17}$ = _____   37. $\frac{7}{21}$ = _____   38. $\frac{1}{19}$ = _____   39. $\frac{16}{23}$ = _____   40. $\frac{4}{7}$ = _____

41. $\frac{9}{11}$ = _____   42. $\frac{18}{21}$ = _____   43. $\frac{19}{23}$ = _____   44. $\frac{1}{13}$ = _____   45. $\frac{10}{17}$ = _____

46. $\frac{3}{23}$ = _____   47. $\frac{4}{17}$ = _____   48. $\frac{17}{21}$ = _____   49. $\frac{6}{19}$ = _____   50. $\frac{6}{21}$ = _____

Gaurav Tekriwal

Convert to decimals.

1. 30 % = ____ 2. 18 % = ____ 3. 55 % = ____ 4. 9 % = ____ 5. 22 % = ___

6. 40 % = ____ 7. 17 % = ____ 8. 6 % = ____ 9. 5 % = ____ 10. 74 % = ___

11. 66 % = ____ 12. 94 % = ____ 13. 44 % = ____ 14. 96 % = ____ 15. 24 % = ___

16. 20 % = ____ 17. 72 % = ____ 18. 64 % = ____ 19. 48 % = ____ 20. 29 % = ___

21. 70 % = ____ 22. 73 % = ____ 23. 75 % = ____ 24. 83 % = ____ 25. 93 % = ___

26. 12 % = ____ 27. 4 % = ____ 28. 28 % = ____ 29. 14 % = ____ 30. 53 % = ___

31. 81 % = ____ 32. 85 % = ____ 33. 97 % = ____ 34. 25 % = ____ 35. 86 % = ___

36. 51 % = ____ 37. 61 % = ____ 38. 77 % = ____ 39. 34 % = ____ 40. 71 % = ___

41. 79 % = ____ 42. 45 % = ____ 43. 11 % = ____ 44. 65 % = ____ 45. 36 % = ___

46. 47 % = ____ 47. 57 % = ____ 48. 95 % = ____ 49. 1 % = ____ 50. 35 % = ___

Calculate the given percent of each value.

1. 55% of 9 = _____   2. 73% of 4 = _____   3. 26% of 6 = _____

4. 66% of 2 = _____   5. 51% of 55 = _____   6. 58% of 398 = _____

7. 98% of 602 = _____   8. 28% of 7 = _____   9. 27% of 974 = _____

10. 72% of 928 = _____   11. 52% of 4 = _____   12. 78% of 9 = _____

13. 33% of 4 = _____   14. 97% of 206 = _____   15. 61% of 964 = _____

16. 56% of 483 = _____   17. 66% of 4 = _____   18. 11% of 220 = _____

19. 64% of 7 = _____   20. 66% of 6 = _____   21. 53% of 48 = _____

22. 10% of 5 = _____   23. 23% of 7 = _____   24. 64% of 735 = _____

25. 87% of 6 = _____   26. 10% of 9 = _____   27. 33% of 585 = _____

28. 11% of 90 = _____   29. 47% of 312 = _____   30. 94% of 8 = _____

31. 44% of 13 = _____   32. 89% of 58 = _____   33. 58% of 589 = _____

34. 55% of 5 = _____   35. 73% of 963 = _____   36. 26% of 90 = _____

37. 66% of 362 = _____   38. 51% of 76 = _____   39. 58% of 6 = _____

40. 98% of 414 = _____   41. 28% of 32 = _____   42. 27% of 148 = _____

43. 72% of 841 = _____   44. 52% of 79 = _____   45. 78% of 969 = _____

46. 33% of 1 = _____   47. 97% of 42 = _____   48. 61% of 5 = _____

49. 56% of 7 = _____   50. 66% of 634 = _____

Gaurav Tekriwal

Provide the conversions for each ratio.

1.

	Ratio	Fraction	Per cent	Decimal
a.		5/6		
b.		2/4		
c.		5/5		
d.		1/4		
e.			66.7%	
f.			33.3%	
g.	8:9			
h.				0.8
i.		6/7		
j.	5:10			

2.

	Ratio	Fraction	Per cent	Decimal
a.		1/1		
b.	3:5			
c.		3/7		
d.		1/4		
e.		2/4		
f.				0.143
g.		6/9		
h.		4/9		
i.			90%	
j.		5/10		

3.

	Ratio	Fraction	Per cent	Decimal
a.			71.4%	
b.		1/4		
c.		5/9		
d.			87.5%	
e.				0.5
f.				0.667
g.	2:3			
h.			100%	
i.	4:5			
j.		6/7		

4.

	Ratio	Fraction	Per cent	Decimal
a.		4/10		
b.		4/6		
c.				0.6
d.				0.333
e.	5:8			
f.		4/4		
g.				0.9
h.			66.7%	
i.				0.5
j.	2:4			

Find if the dividend is divisible by the divisor or not.

1.
$77\overline{)2,387}$

2.
$96\overline{)1,248}$

3.
$68\overline{)4,488}$

4.
$40\overline{)3,040}$

5.
$25\overline{)1,750}$

6.
$46\overline{)2,254}$

7.
$80\overline{)5,120}$

8.
$38\overline{)2,546}$

9.
$19\overline{)1,178}$

10.
$92\overline{)5,244}$

11.
$53\overline{)636}$

12.
$68\overline{)6,460}$

13.
$57\overline{)5,130}$

14.
$61\overline{)2,074}$

15.
$69\overline{)4,002}$

16.
$19\overline{)437}$

17.
$13\overline{)156}$

18.
$69\overline{)5,313}$

19.
$91\overline{)6,552}$

20.
$78\overline{)4,524}$

21.
$53\overline{)2,438}$

22.
$47\overline{)1,269}$

23.
$29\overline{)1,450}$

24.
$61\overline{)2,196}$

25.
$85\overline{)3,825}$

26.
$35\overline{)490}$

27.
$90\overline{)2,700}$

28.
$78\overline{)5,928}$

29.
$62\overline{)744}$

30.
$56\overline{)4,256}$

31.
$16\overline{)272}$

32.
$18\overline{)1,188}$

33.
$47\overline{)3,713}$

34.
$95\overline{)5,605}$

35.
$75\overline{)3,450}$

36.
$83\overline{)3,735}$

37.
$15\overline{)270}$

38.
$65\overline{)3,380}$

39.
$78\overline{)1,404}$

40.
$41\overline{)1,435}$

41.
$48\overline{)1,680}$

42.
$34\overline{)544}$

43.
$32\overline{)2,624}$

44.
$33\overline{)3,036}$

45.
$26\overline{)2,340}$

46.
$27\overline{)351}$

47.
$53\overline{)2,491}$

48.
$40\overline{)3,200}$

49.
$65\overline{)6,045}$

50.
$14\overline{)1,120}$

Convert the values.

1. $92^2 =$ _____ 2. $9^2 =$ _____ 3. $53^2 =$ _____ 4. $8^2 =$ _____ 5. $5^2 =$ _____

6. $12^2 =$ _____ 7. $4^2 =$ _____ 8. $7^2 =$ _____ 9. $6^2 =$ _____ 10. $\overline{20}^2 =$ _____

11. $52^2 =$ _____ 12. $3^2 =$ _____ 13. $70^2 =$ _____ 14. $75^2 =$ _____ 15. $56^2 =$ _____

16. $59^2 =$ _____ 17. $64^2 =$ _____ 18. $85^2 =$ _____ 19. $17^2 =$ _____ 20. $55^2 =$ _____

21. $15^2 =$ _____ 22. $99^2 =$ _____ 23. $80^2 =$ _____ 24. $2^2 =$ _____ 25. $76^2 =$ _____

26. $79^2 =$ _____ 27. $1^2 =$ _____ 28. $63^2 =$ _____ 29. $30^2 =$ _____ 30. $62^2 =$ _____

31. $95^2 =$ _____ 32. $50^2 =$ _____ 33. $58^2 =$ _____ 34. $78^2 =$ _____ 35. $67^2 =$ _____

36. $37^2 =$ _____ 37. $87^2 =$ _____ 38. $66^2 =$ _____ 39. $27^2 =$ _____ 40. $31^2 =$ _____

41. $10^2 =$ _____ 42. $90^2 =$ _____ 43. $19^2 =$ _____ 44. $82^2 =$ _____ 45. $88^2 =$ _____

46. $16^2 =$ _____ 47. $48^2 =$ _____ 48. $43^2 =$ _____ 49. $14^2 =$ _____ 50. $60^2 =$ _____

Convert the values.

1. $638^2 =$ _____
2. $990^2 =$ _____
3. $218^2 =$ _____
4. $373^2 =$ _____

5. $404^2 =$ _____
6. $426^2 =$ _____
7. $328^2 =$ _____
8. $392^2 =$ _____

9. $758^2 =$ _____
10. $298^2 =$ _____
11. $108^2 =$ _____
12. $796^2 =$ _____

13. $873^2 =$ _____
14. $416^2 =$ _____
15. $260^2 =$ _____
16. $242^2 =$ _____

17. $527^2 =$ _____
18. $467^2 =$ _____
19. $179^2 =$ _____
20. $217^2 =$ _____

21. $862^2 =$ _____
22. $559^2 =$ _____
23. $696^2 =$ _____
24. $736^2 =$ _____

25. $281^2 =$ _____
26. $987^2 =$ _____
27. $159^2 =$ _____
28. $907^2 =$ _____

29. $529^2 =$ _____
30. $949^2 =$ _____
31. $269^2 =$ _____
32. $875^2 =$ _____

33. $888^2 =$ _____
34. $475^2 =$ _____
35. $358^2 =$ _____
36. $393^2 =$ _____

37. $751^2 =$ _____
38. $306^2 =$ _____
39. $936^2 =$ _____
40. $610^2 =$ _____

41. $625^2 =$ _____
42. $157^2 =$ _____
43. $810^2 =$ _____
44. $596^2 =$ _____

45. $303^2 =$ _____
46. $284^2 =$ _____
47. $900^2 =$ _____
48. $112^2 =$ _____

49. $521^2 =$ _____
50. $779^2 =$ _____

Convert the values.

1. 56^3 = _____   2. 25^3 = _____   3. 4^3 = _____   4. 8^3 = _____

5. 5^3 = _____   6. 44^3 = _____   7. 1^3 = _____   8. 7^3 = _____

9. 9^3 = _____   10. 34^3 = _____   11. 18^3 = _____   12. 45^3 = _____

13. 17^3 = _____   14. 67^3 = _____   15. 77^3 = _____   16. 15^3 = _____

17. 52^3 = _____   18. 6^3 = _____   19. 2^3 = _____   20. 49^3 = _____

21. 23^3 = _____   22. 72^3 = _____   23. 37^3 = _____   24. 36^3 = _____

25. 69^3 = _____   26. 13^3 = _____   27. 97^3 = _____   28. 81^3 = _____

29. 98^3 = _____   30. 3^3 = _____   31. 32^3 = _____   32. 70^3 = _____

33. 43^3 = _____   34. 31^3 = _____   35. 80^3 = _____   36. 87^3 = _____

37. 99^3 = _____   38. 38^3 = _____   39. 30^3 = _____   40. 62^3 = _____

41. 29^3 = _____   42. 93^3 = _____   43. 85^3 = _____   44. 94^3 = _____

45. 57^3 = _____   46. 54^3 = _____   47. 90^3 = _____   48. 16^3 = _____

49. 22^3 = _____   50. 73^3 = _____

Calculate the root of each value.

1. $\sqrt{7,056}$ = _____   2. $\sqrt{1,225}$ = _____   3. $\sqrt{5,184}$ = _____   4. $\sqrt{7,396}$ = _____

5. $\sqrt{2,401}$ = _____   6. $\sqrt{8,649}$ = _____   7. $\sqrt{5,476}$ = _____   8. $\sqrt{1,089}$ = _____

9. $\sqrt{9,025}$ = _____   10. $\sqrt{1,681}$ = _____   11. $\sqrt{3,249}$ = _____   12. $\sqrt{1,296}$ = _____

13. $\sqrt{2,500}$ = _____   14. $\sqrt{2,916}$ = _____   15. $\sqrt{3,481}$ = _____   16. $\sqrt{6,561}$ = _____

17. $\sqrt{2,304}$ = _____   18. $\sqrt{7,569}$ = _____   19. $\sqrt{4,900}$ = _____   20. $\sqrt{10,000}$ = _____

21. $\sqrt{4,225}$ = _____   22. $\sqrt{1,849}$ = _____   23. $\sqrt{9,409}$ = _____   24. $\sqrt{5,625}$ = _____

25. $\sqrt{4,624}$ = _____   26. $\sqrt{2,601}$ = _____   27. $\sqrt{6,400}$ = _____   28. $\sqrt{4,489}$ = _____

29. $\sqrt{8,836}$ = _____   30. $\sqrt{5,929}$ = _____   31. $\sqrt{7,921}$ = _____   32. $\sqrt{8,464}$ = _____

33. $\sqrt{1,369}$ = _____   34. $\sqrt{3,844}$ = _____   35. $\sqrt{7,744}$ = _____   36. $\sqrt{9,216}$ = _____

37. $\sqrt{3,600}$ = _____   38. $\sqrt{2,704}$ = _____   39. $\sqrt{1,764}$ = _____   40. $\sqrt{1,444}$ = _____

41. $\sqrt{3,721}$ = _____   42. $\sqrt{3,025}$ = _____   43. $\sqrt{4,356}$ = _____   44. $\sqrt{9,801}$ = _____

45. $\sqrt{5,329}$ = _____   46. $\sqrt{1,936}$ = _____   47. $\sqrt{3,136}$ = _____   48. $\sqrt{5,041}$ = _____

49. $\sqrt{3,364}$ = _____   50. $\sqrt{6,241}$ = _____

Calculate the root of each value correct to two decimal places.

1. $\sqrt{2,747}$ = _____ 2. $\sqrt{2,734}$ = _____ 3. $\sqrt{1,249}$ = _____

4. $\sqrt{3,541}$ = _____ 5. $\sqrt{3,780}$ = _____ 6. $\sqrt{6,385}$ = _____

7. $\sqrt{6,365}$ = _____ 8. $\sqrt{9,218}$ = _____ 9. $\sqrt{5,029}$ = _____

10. $\sqrt{4,115}$ = _____ 11. $\sqrt{1,559}$ = _____ 12. $\sqrt{5,703}$ = _____

13. $\sqrt{4,079}$ = _____ 14. $\sqrt{4,046}$ = _____ 15. $\sqrt{9,518}$ = _____

16. $\sqrt{5,702}$ = _____ 17. $\sqrt{3,963}$ = _____ 18. $\sqrt{6,024}$ = _____

19. $\sqrt{3,251}$ = _____ 20. $\sqrt{5,617}$ = _____ 21. $\sqrt{6,012}$ = _____

22. $\sqrt{8,501}$ = _____ 23. $\sqrt{1,672}$ = _____ 24. $\sqrt{9,154}$ = _____

25. $\sqrt{2,140}$ = _____ 26. $\sqrt{6,763}$ = _____ 27. $\sqrt{3,466}$ = _____

28. $\sqrt{5,544}$ = _____ 29. $\sqrt{4,438}$ = _____ 30. $\sqrt{7,048}$ = _____

31. $\sqrt{3,065}$ = _____ 32. $\sqrt{4,006}$ = _____ 33. $\sqrt{9,038}$ = _____

34. $\sqrt{2,905}$ = _____ 35. $\sqrt{9,270}$ = _____ 36. $\sqrt{1,081}$ = _____

37. $\sqrt{6,150}$ = _____ 38. $\sqrt{7,316}$ = _____ 39. $\sqrt{2,493}$ = _____

40. $\sqrt{7,452}$ = _____ 41. $\sqrt{1,636}$ = _____ 42. $\sqrt{3,846}$ = _____

43. $\sqrt{3,654}$ = _____ 44. $\sqrt{3,602}$ = _____ 45. $\sqrt{3,539}$ = _____

46. $\sqrt{9,980}$ = _____ 47. $\sqrt{3,534}$ = _____ 48. $\sqrt{8,623}$ = _____

49. $\sqrt{1,561}$ = _____ 50. $\sqrt{3,639}$ = _____

Calculate the cube root of each value.

1. $\sqrt[3]{614,125}$ = _____

2. $\sqrt[3]{405,224}$ = _____

3. $\sqrt[3]{830,584}$ = _____

4. $\sqrt[3]{753,571}$ = _____

5. $\sqrt[3]{166,375}$ = _____

6. $\sqrt[3]{373,248}$ = _____

7. $\sqrt[3]{704,969}$ = _____

8. $\sqrt[3]{884,736}$ = _____

9. $\sqrt[3]{636,056}$ = _____

10. $\sqrt[3]{140,608}$ = _____

11. $\sqrt[3]{97,336}$ = _____

12. $\sqrt[3]{216,000}$ = _____

13. $\sqrt[3]{474,552}$ = _____

14. $\sqrt[3]{175,616}$ = _____

15. $\sqrt[3]{117,649}$ = _____

16. $\sqrt[3]{328,509}$ = _____

17. $\sqrt[3]{314,432}$ = _____

18. $\sqrt[3]{512,000}$ = _____

19. $\sqrt[3]{912,673}$ = _____

20. $\sqrt[3]{148,877}$ = _____

21. $\sqrt[3]{592,704}$ = _____

22. $\sqrt[3]{729,000}$ = _____

23. $\sqrt[3]{778,688}$ = _____

24. $\sqrt[3]{132,651}$ = _____

25. $\sqrt[3]{658,503}$ = _____

26. $\sqrt[3]{438,976}$ = _____

27. $\sqrt[3]{456,533}$ = _____

28. $\sqrt[3]{571,787}$ = _____

29. $\sqrt[3]{103,823}$ = _____

30. $\sqrt[3]{238,328}$ = _____

31. $\sqrt[3]{262,144}$ = _____

32. $\sqrt[3]{125,000}$ = _____

33. $\sqrt[3]{343,000}$ = _____

34. $\sqrt[3]{274,625}$ = _____

35. $\sqrt[3]{300,763}$ = _____

36. $\sqrt[3]{804,357}$ = _____

37. $\sqrt[3]{493,039}$ = _____

38. $\sqrt[3]{389,017}$ = _____

39. $\sqrt[3]{287,496}$ = _____

40. $\sqrt[3]{531,441}$ = _____

41. $\sqrt[3]{970,299}$ = _____

42. $\sqrt[3]{941,192}$ = _____

43. $\sqrt[3]{157,464}$ = _____

44. $\sqrt[3]{1,000,000}$ = _____

45. $\sqrt[3]{226,981}$ = _____

46. $\sqrt[3]{681,472}$ = _____

47. $\sqrt[3]{110,592}$ = _____

48. $\sqrt[3]{185,193}$ = _____

49. $\sqrt[3]{857,375}$ = _____

50. $\sqrt[3]{195,112}$ = _____

Convert the values.

1. $2^4 =$ _____
2. $9^4 =$ _____
3. $59^4 =$ _____

4. $32^4 =$ _____
5. $4^4 =$ _____
6. $20^4 =$ _____

7. $23^4 =$ _____
8. $3^4 =$ _____
9. $49^4 =$ _____

10. $60^4 =$ _____
11. $50^4 =$ _____
12. $63^4 =$ _____

13. $87^4 =$ _____
14. $8^4 =$ _____
15. $1^4 =$ _____

16. $47^4 =$ _____
17. $6^4 =$ _____
18. $7^4 =$ _____

19. $35^4 =$ _____
20. $62^4 =$ _____
21. $89^4 =$ _____

22. $42^4 =$ _____
23. $22^4 =$ _____
24. $70^4 =$ _____

25. $51^4 =$ _____
26. $31^4 =$ _____
27. $48^4 =$ _____

28. $64^4 =$ _____
29. $68^4 =$ _____
30. $5^4 =$ _____

31. $82^4 =$ _____
32. $34^4 =$ _____
33. $11^4 =$ _____

34. $28^4 =$ _____
35. $95^4 =$ _____
36. $66^4 =$ _____

37. $33^4 =$ _____
38. $55^4 =$ _____
39. $77^4 =$ _____

40. $18^4 =$ _____
41. $61^4 =$ _____
42. $14^4 =$ _____

43. $39^4 =$ _____
44. $74^4 =$ _____
45. $96^4 =$ _____

46. $38^4 =$ _____
47. $76^4 =$ _____
48. $85^4 =$ _____

49. $83^4 =$ _____
50. $57^4 =$ _____

Find the product.

1. $(-3u+5v)(-3u+8v)$

2. $(-5x+y)(-3x-7y)$

3. $(-a-8b)(2a+7b)$

4. $(-4m-6n)(-m-8n)$

5. $(6x+5y)(-6x+7y)$

6. $(-u-8v)(7u+v)$

7. $(5x+y)(3x-6y)$

8. $(6x+5y)(7x-3y)$

9. $(-5x-2y)(7x+5y)$

10. $(2a+6b)(5a+b)$

11. $(x+3y)(-3x-4y)$

12. $(-6a-2b)(-a-6b)$

13. $(2a+3b)(-4a+6b)$

14. $(-5a-7b)(-6a-7b)$

15. $(2x-3y)(-x-4y)$

16. $(6u-v)(-7u+7v)$

17. $(-u-3v)(-8u-v)$

18. $(8x-7y)(-x-2y)$

19. $(8x+3y)(7x-y)$

20. $(-2x+8y)(-7x+8y)$

21. $(m+4n)(-6m+3n)$

22. $(-2x-2y)(2x-3y)$

23. $(2x+6y)(8x-5y)$

24. $(8u-4v)(8u+v)$

25. $(-3a-4b)(-a-8b)$

26. $(8x+8y)(-x-5y)$

27. $(8x-8y)(-6x-3y)$

28. $(2x+4y)(-4x+2y)$

29. $(5x+7y)(4x-5y)$

30. $(4a-7b)(4a-5b)$

31. $(-5a-3b)(2a+5b)$

32. $(-3x-8y)(-7x-4y)$

33. $(7u+2v)(-6u-v)$

34. $(x-y)(-x-8y)$

35. $(3x-3y)(-5x+2y)$

36. $(6x-8y)(-x-7y)$

37. $(5x+3y)(8x-7y)$

38. $(-a+4b)(-4a-4b)$

39. $(-2u-5v)(3u-4v)$

40. $(7x-y)(5x+8y)$

41. $(2x+5y)(7x-7y)$

42. $(6x-6y)(x+6y)$

43. $(4a-b)(a+3b)$

44. $(-8x-2y)(-2x+5y)$

45. $(3x+5y)(-7x+4y)$

46. $(-u-2v)(7u+2v)$

47. $(-3x-8y)(-8x+y)$

48. $(-3a+2b)(7a-3b)$

49. $(4a+8b)(-4a-5b)$

50. $(-5x-2y)(7x-8y)$

Simplify each expression.

1. $(6n^3 + 3n) - (3n + 5n^3) + (5n^3 - 6n)$

2) $(5 + 4k^3) + (k - 1) + (1 - 5k^3)$

3. $(5x3 + 2x^2) - (3x + 8x^3) + (6x^2 + 8x^3)$

4. $(6x^3 - 6x) + (4x - 4x^3) - (x - x^3)$

5. $(8 - 3n^2) + (8n^4 - n) - (4 - 6n^2)$

6. $(3n^4 - 5) - (2n - 3n^4) + (3 - 8n^4)$

7. $(3 - n^3) - (5n + 6) + (3 + 4n)$

8. $(5x^3 + 6) - (x^3 - 2) + (x^3 + 2x^4)$

9) $(7m^4 - 3m^3) + (m^4 + 7m^3) + (8m^4 - 5m^3)$

10. $(6 + 7a^2) - (8 + 3a^2) + (8 - 8a^2)$

11. $(7x^4 - 8x^2) + (x^3 + 7x^2) - (7x^3 + 4x^4)$

12. $(6x^3 - 7x) - (3x + 2x^3) + (2x^3 + 5x)$

13. $(6 - 4x^2) - (3 - 6x^2) + (5x^2 - 5)$

14. $(8n - 5n^3) - (8n^3 + 5n) - (7n + 2n^3)$

15. $(6x + 3x^2) + (x - 5x^2) - (2x - 2x^2)$

16. $(7x^4 + x) - (6x^4 + 6x^2) + (x^2 - 6x)$

17. $(7x^4 - 5x^2) + (6x^4 + 6x^2) - (x^3 - 6x^4)$

18. $(6\,p^4 + 6\,p^3) - (4\,p - 4\,p^3) - (8\,p^2 - p^3)$

19. $(8m - 8m^4) + (4m^4 + 6m) + (4m^4 + 3m)$

20. $(7n^2 - 4n^3) + (3n^2 + 3n^3) + (6n^3 - 7n^2)$

21. $(5 + 4r^2) - (2r^2 + 6) - (6r^2 + 8)$

22. $(3r^2 + 7r^3) + (5r^3 + 4r^2) + (6r^2 + 6r^3)$

23. $(8 - n^4) + (7 - 5n^4) + (6n^4 + 7)$

24. $(6v^4 + 3v^2) - (v^4 + 2v^2) - (3v^4 + 7)$

25. $(6m + 5m^2) - (m^2 + 2m^4) + (8m + m^4)$

26. $(2 - x^2) + (x^2 + 6) + (4 - 3x^2)$

27. $(6k + 4k^2) - (3k + 3k^4) + (6k^4 + 4k^2)$

28. $(6 + 3n) - (8 + 3n) - (n - 3)$

29. $(4 - 6n^2) - (2n^2 - 6) - (n - 7n^3)$

30. $(n + 2n^2) - (4n - 2n^2) + (5n^2 - 3n)$

31. $(3n^2 - 5n) + (3n^3 - 4n^2) - (5n^2 - 8n^3)$

32. $(1 - 2x) + (8x^3 + x) + (4x^3 - 6x)$

33. $(5n^2 - 8n^3) - (3n^3 - 6n^2) + (5n^2 - 4n^3)$

34. $(4x^3 + 1) + (x^2 + 2x^3) + (6x^3 + 5x^2)$

35. $(7x^2 + 3) + (8x^2 - 7) - (7x^2 + 3)$

36. $(n^4 - 2) + (3n^4 - 5) + (4n^4 - 1)$

37. $(5 - 4r^4) + (7r^4 + 5) - (r^4 + 5)$

38. $(8k^4 + 4k) + (8k^4 - 2k) - (k^4 + 3k)$

39. $(4m - 2m^2) + (8 + 3m^2) + (7 - 4m^4)$

40. $(6x^2 - 5x) + (8x - 7x^2) - (5x + 7x^2)$

41. $(5v^4 + 3) - (7 - 8v^4) + (v^4 + 1)$

42. $(5 - 5x^3) + (4 - 4x^3) + (8 + 8x^3)$

43. $(n^4 - 7n^2) + (4n^2 - 2n^4) + (4n^2 - 5)$

44. $(2\,p^3 + 4\,p^2) - (2\,p^3 + 5) - (3\,p^2 - 8)$

45. $(k^3 + 6) + (4 - 2k^3) + (7 - k^3)$

46. $(3a^2 + a^4) + (3a^2 + 5a^4) - (8a^4 + 4a^2)$

47. $(8b^4 - 7) - (5b^2 - 8) + (5 - 5b^4)$

48. $(2n + n^3) - (6n - 6n^3) - (6n^3 + 6n)$

49. $(8x^2 + 4x) - (2x - 6x^2) - (3x^2 + 2x)$

50. $(3 - 7k^2) - (7k^2 - 7k) - (7k^2 - 2k)$

Solve for x and y.

1. $x + 3y = 3$
    $4x + 3y = -15$

2. $7x + 2y = 10$
    $x - y = 4$

3. $7x - 4y = -28$
    $x - 2y = 6$

4. $x - 3y = -24$
    $14x - 3y = 15$

5. $x + y = -3$
    $x - 2y = -12$

6. $x + 2y = 8$
    $2x + y = -2$

7. $9x - 8y = -8$
    $x - 8y = 56$

8. $5x - 8y = -8$
    $3x + 8y = -56$

9. $7x + y = -6$
    $x - 2y = -18$

10. $x - 2y = -4$
    $4x + y = -7$

11. $2x + 7y = -49$
    $2x - 7y = 21$

12. $3x - y = -6$
    $4x + y = -1$

13. $x - 2y = 10$
    $7x + 8y = 48$

14. $5x - 9y = -63$
    $8x + 9y = -54$

15. $9x - 5y = -5$
    $x + 5y = -45$

16. $5x + 2y = -10$
    $x + 4y = 16$

17. $x - 2y = 14$
    $6x + y = 6$

18. $2x + 7y = 14$
    $13x + 7y = -63$

19. $5x - 3y = 18$
    $8x + 3y = 21$

20. $x - 2y = 4$
    $13x - 6y = -48$

21. $13x + 8y = -56$
    $x = -8$

22. $5x - 2y = -14$
    $x - 6y = 42$

23. $4x + 3y = -21$
    $x - 2y = -8$

24. $2x - 7y = -63$
    $5x + 7y = 14$

25. $y = -9$
    $17x + 2y = 16$

26. $12x - 5y = -25$
    $y = -7$

27. $x = 6$
    $5x + 3y = 18$

28. $3x - 7y = -63$
    $11x + 7y = -35$

29. $x + 4y = -28$
    $5x - 2y = -8$

30. $15x - 4y = 36$
    $x - 4y = -20$

31. $13x + y = 9$
    $2x - y = 6$

32. $2x + y = 1$
    $2x + 5y = -35$

33. $2x - 3y = 9$
    $5x - 2y = -16$

34. $9x - 4y = 24$
    $3x + 4y = 24$

35. $x + 9y = 63$
    $5x - 3y = 27$

36. $x - 4y = 32$
    $17x - 4y = -32$

37. $4x - 3y = -15$
    $2x + 3y = -21$

38. $x + 3y = -18$
    $4x + 3y = 9$

39. $2x - y = -9$
    $2x - 9y = 63$

40. $x + 4y = 4$
    $9x + 4y = -28$

41. $4x + y = -7$
    $x + y = 2$

42. $x + 8y = 64$
    $9x - 8y = 16$

43. $7x - 9y = 54$
    $x + 9y = 18$

44. $2x + y = 2$
    $x - 3y = -27$

45. $4x + y = 5$
    $4x - y = 3$

46. $x - 6y = 30$
    $2x + y = 8$

47. $5x - 9y = 18$
    $x + 9y = 36$

48. $6x - y = 7$
    $x - y = -3$

49. $x + y = -6$
    $6x - y = -1$

50. $8x - y = 8$
    $x - y = -6$

Solve the following.

1. $2m^2 + 2m - 12 = 0$ 2. $2x^2 - 3x - 5 = 0$

3. $x^2 + 4x + 3 = 0$ 4. $2x^2 + 3x - 20 = 0$

5. $4b^2 + 8b + 7 = 4$ 6. $2m^2 - 7m - 13 = -10$

7. $2k^2 + 9k = -7$ 8. $5r^2 = 80$

9. $2x^2 - 36 = x$ 10. $5x^2 + 9x = -4$

Acknowledgements

I wish to express my gratitude to my guru, His Holiness Jagadguru Shankaracharya Shri Nischalananda Saraswatiji of Govardhan Matha, Puri, Odisha, for his blessings, guidance and inspiration.

I would like to thank all the people who saw me through this book, particularly my parents who showered me with unconditional love and provided constant encouragement, and my wife, Shree, for her undying faith in me and for being such a strong critic of my work. And of course, Miraaya, my energetic one-year-old, has to be thanked for allowing me to work during her playtime.

I would also like to thank my friends Varun Poddar and Abhishek Agarwal for the stimulating discussions on making *Maths Sutra* a global brand.

Many thanks to Pictofigo.com for supplying illustrations for the book on time.

I am grateful to Vaishali Mathur at Penguin Random House for believing in this book, and to all those at PRH who lent support, offered suggestions and assisted in the making of *Maths Sutra*.